Grace Lau's Hydraulic Engineering School Papers

Grace Lau

Copyright © 2014 Grace H. Lau

All rights reserved.

DEDICATION

To my father who mentored me during civil engineering school

CONTENTS

Bottled Water: Consumer Opinion

Prepared by: **Grace Lau**

for

HLSC 114: Health Ecology and Health

California State University, Sacramento

Report Due on September 9, 2011

Sales of bottled water have increased dramatically in recent years. According to Priesnitz (2007), there was a 57 percent increase in the global consumption of bottled water between 2000 to 2004. The global consumption of bottled water reached 154 billion liters in 2004 compared to 98 billion liters in 2000. According to Arnold (2006), some of the largest increases in bottled water consumption occurred in developing countries. Arnold (2006) found that the United States is the world's leading consumer of bottled water, and the consumption per person was increasing by 44 to 50 percent between 1999 and 2004 in the United States. The trend has generated some studies to understand consumer's preferences for bottled water and the qualities of bottled water.

Public perception that bottled water pure and safe promotes the sale. According to Ward, Cain, Mullally, Holliday (2009), most participates in her study believed bottled water has added health values. However, there is no sufficient data or research to support the claim. Instead, Kokkinakis (2011) stated that bottled water may contain heterotropic plate count (HPC) bacteria during bottling and storage at ambient temperatures. Currently bottled water does not to provide full disclosure of all test results for all contaminants to consumers.

Another reason for consumption of bottled water is some people are not satisfied with the smell, taste, and odor of tap water. Doria (2006) found that a study on exploring consumers' preferences of bottled water versus tap water reviewed that consumers in USA, Canada, and France buy bottled water because they are not satisfied with the organoleptics of the tap water (the taste, odor, sight). The finding is supported by other studies. A study from Levallois, Groundin, and Gingras (1999) uses questionnaires from Canadian regions to show that respondents identified organoleptics as the main reason for drinking bottled water from 63-80%, depending on the region.

Bottled water is convenient and useful during emergencies and other situation. According to Symons (1997) during emergencies bottled water can be a vital source of drinking water for people without water. The Federal Emergency Management Agency (FEMA) stated that on July 3, 2007 FEMA has delivered 13 truck loads of bottled water to the victims of Kansas floods. Since bottled water is convenient and portable, and people can buy it anywhere and use it any time, Americans can access bottled water at the gym, at work, after a workout, at just about any time. Priesnitz (2007) says that people can take bottled water to work where they can drink it anywhere.

Bottled water is a substitute for alcoholic and traditional soft drinks and it is a part of pop culture. Priesnitz (2007) says bottled water is a nearly ideal consumer product: it is healthy, hypoallergenic, calorie free, and contains no artificial colors, flavors, and trans fats. It is also non-addictive, and caffeine-free. It is consumed increasingly by more Americans, and it has its place in commercial value. Americans are prepared to pay what they perceived to be purer or healthier product in degraded or contaminated environments.

Bottled water can provide and serve as alternative where municipal water sources are compromised. According to Arnold (2006), bottled water can provide clean water where local resources are not there, but bottled water is not the answer in the developed world, nor does it solve problems for the 1.1 billion people who lack a secure water supply. Doria (2006) reports that bottled water is more popular where communities have poor tap water sources.

Bottled water can be a choice to those who look beyond the drinking tap water. When there is a disease outbreak or the tap water is contaminated, more people will buy bottled water because they lose trust in tap water. According to Doria (2006), suspicion and a history of past problems may happen during major catastrophes can impact the public's behavior. Lonnon (2004) reports an instance where this happened was in 1998, the Syndey Cryptosporidium and Giardia outbreak helped the Australian bottled water market. In emergency, bottled water can be an absolutely critical lifesaver.

While bottled water has its value, there is recent concern about the negative impact of bottle water on our ecology and health. Bottled water is not environmentally friendly. It costs extra money and energy to package bottled water. It requires large amounts of energy to produce and distribute. The energy used each year in making the bottles needed to meet the demand for bottled water in the United States is equivalent to more than 17 million barrels of oil. According to Doria (2006), that's enough to fuel over one million cars for a year. Most of the bottles are made of the oil-derived polyethylene (PET). Priesnitz stated that the production of the PET generates 100 times toxic emissions than the production of the same amount of glass, and many of the plastic water bottles are discarded in the landfills after the water is consumed. Both factors exacerbate the environmental problem.

Bottled water can be contaminated because it is stored in a plastic bottle or container. Various studies have found that some bottled water contains water contamination. Dora stated that traces of alkylphenols and phthalates were found in some bottled waters. The chemical will interfere with the hormone system in human and animals that are known to cause tumors and birth defects and developmental disorders. According to Olson, Leiba, Sharp, Houlihan (1999) some cancer chemical contaminants were found in some bottled water. Priesnitz (2007) stated that some bottled water were contaminated at levels violating strict state (California) limits.

Bottled water has limited shelf life and it does not

store well, because it generally does not contain a disinfectant. Symons stated that in the United States only some states require bottled water's shelf life to be stamped on the label. For example in New York the expiration date is two years from the date of bottling. The water is presumably safe to consume within the expiration date. However, according to Symons (1997) studies have shown that microbe may grow in the bottles while on grocers' shelves. According to Jackson (2008), The FDA requires bottlers to regularly test for contaminants, but the agency considers bottled water a low-risk product and most plants are not inspected for contaminants annually.

Bottled water is not being regulated as tight as the tap water in America. The bottle water are not regulated by the Safe Drinking Water Act. According to Symons

(2009), Federal and state regulation require the accuracy of labels or claims of purity, but the quality of the finished product is not government monitored. In the U.S. bottled water should be tested for bacteria weekly, but the municipal water is mandated to test for bacteria every four hours. The FDA needs to make its rules stricter for bottled water. Olson (1999) states that the FDA's rules completely exempt 60-70 percent of the bottled water sold in the United States from the agency's bottled water standards, because the FDA says that its rules do not apply to water packaged and sold within the same state.

Bottled water is expensive because bottled water companies have to pay for heavy advertisement, filtering, and bottling the water. In addition there are cost for packaging plus shipment for bottled water.

According to Doria (2006), bottled water costs 1,000 times more than tap water. Dora stated the cost for bottled water is as much as $5 for a gallon or $1.25 per liter. This is much pricier than the cost of a gallon of a gasoline.

Some bottled water labels are misleading to consumers even though the Food and Drug Administration has established rules for the labeling of bottled water. According to Olson (1999) deceptive bottled water labeling was a widespread practice, and the use of descriptive terminology to suggest extraordinary quality of bottled water was pretty common. This can create hazard for people who have special health issues and are weak in their immune system. For example some bottled water labeled sodium-free contains some sodium, maybe too much for those people who are on highly restricted sodium diet.

Bottled water has its value as an alternative for emergency use. It is great to have bottled water available for natural disasters or other times when the public supply might not be available or might be susceptible to contamination. In the long run, I recommend the bottled water subject to the same strict regulatory standards as the drinking tap water. FDA should set strict limits for contaminants of concern in bottled water including arsenic, hetertrophic-plate-count bacteria, E.coli, Pseudomonas aeruginosa, and synthetic organic chemicals. FDA rules should apply to all bottles water distributed nationally or within a state. FDA should require water bottle labels to disclose contaminants, the exact water source, treatment, and other key information as is now required of the tap water system. If Americans choose to buy bottled water, they deserve to know that it is safe, and they have the right to know what is in it.

References

1. Arnold, E. & Larsen, J. (2006). Providing a plan to save civilization. *Earth Policy Institute*, Retrieved from

   http://www.earth-policy.org/plan_b_updates/2006/update51

2. Bicking, C., Merkel, L., Sekhar D. (2011). Parents Perceptions of Water Safety and Quality. *Journal of Community Health,* 33, 5. doi 10.1007/s10900-011-9436-9

3. Doria, M. (2006) Bottled Water Versus Tap Water: Understanding Consumers' Preferences. *Journal of Water and Health*. 4, 271-276. doi:10.2166/wh.2006.008

4. Federal Emergency Management Agency. (2007). *FEMA/state and other federal agencies continue to assist those affected by Kansas foods* (Release number:1711- 004). Washington, D.C.: U.S. Government Printing Office.

5. H.H. Lai (personal communication, August 8, 2011)

6. Houlihan, J., Leiba, N., Naidenko, O., Sharp, R.

(2008). Bottled Water Quality Investigation: 10 Major Brands, 38 Pollutants; Bottled water contains disinfection byproducts, fertilizer residue, and pain medication. EWG Research. 1, 1-8 Retrieved from: http://www.ewg.org/reports/BottledWater/Bottled-Water- Quality-Investigation

7. Jackson, L. (April 2008) Environmental Protection Agency. Factoids: drinking water and ground water statistics for 2007. March 2008, April 2008. Retrieved from: http://www.epa.gov/safewater/data/getdata.html

8. Jamerson, M., Kaneshiro, E., Marciano-Cabral, F. (2010). Free-living amoebae, Leionella and Mycobacterium in tap water supplied by a municipal drinking water utility in the USA. *Journal of Water and Health* 8, 71-82. doi: 10.2166/wh.2009.129

9. Kokkinakis, E. (2011) Heterotrophic Bacteria in Bottled Water. Food Control 19, 957–961. doi:10.1016/j.foodcont.2007.10.001

10. Levallois, P., Grondin, J. & Gingras, S. (1999) Evaluation of consumer attitudes on taste and tap water alternatives in Quebec. Water Science Technology. 40, 135- 139. doi: 10.1016/S0273-1223(99)00549-1

11. Lonnon, K. (2004) Bottled Water Drowns the Competition. Retrieved from: http://www.ferret.com.au/articles/98/0c01fc9.asp

12. M Fox. (2009, Oct 8). What's in Bottled Water: What are FDA standards for bottled water? [Web log comment]. Retrieved from http://breakfornews.com/forum/viewtopic.php?t=5403

13. New Study Finds Fault With Some Bottled Waters; Tap Water a Better Bet. (2009). *Environmental Nutrition*, 32, 3. Retrieved from: www.environmentalnutrition.com

14. Olson, E., Poling, D., Solomon, G. (1999, March). Bottled Water Pure Drink or Pure Hype? Retrieved from: http://www.nrdc.org/water/drinking/bw/bwinx.asp

15. Priesnitz, W. (2007) Bottled Water Or Tap Water? *Natural Life*, 114, 8-11. Retrieved from: http://www.naturallifemagazine.com/0704/water.htm

16. Ward, L., Cain, O., Mullally, R., Holliday, K., (2009) Health Beliefs about Bottled Water: a Qualitative Study. *BMC Public Health*, 196. doi:10.1186/1471-2458-9-196

17. Symons, J.M. (1997). Plain Talk About Drinking Water: Questions and Answers About the Water You Drink. Denver, Colorado: American Water Works Association.

18. United Nations Children's Fund and Water Supply and Sanitation Collaborative Council. (2000). *Global Water Supply and Sanitation Assessment 2000 Report* (1st ed.). Washington, DC: Bellarny, C., Brunkfland, G.H.

Open Channel Flow Pedestrian Bridge Hydraulic Modeling Project Using TuFlow Technology and Software Modeling System (SMS) along American River

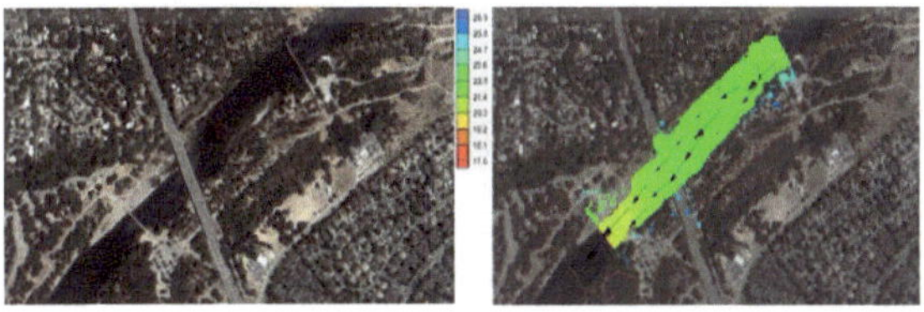

CE 139: Open Channel Hydraulics

Final Project

SACRAMENTO STATE

Jose Dominguez
Grace Lau
Gustavo Landa
Kyle Grijalva

I. Purpose

In 1958, Sunrise Boulevard Bridge was built and has been widened in the past to accommodate for westbound and eastbound vehicular traffic. The problem has been the height of the underside of the Sunrise Boulevard Bridge, which has deteriorated due to a combination of aging, vehicle traffic, and exposure to river water. The downstream pedestrian bridge must be removed to improve hydraulic design conditions. The purpose of this project is to solve the problem by applying open channel flow knowledge acquired during the semester. River two-dimensional models will be created in SMS (software modeling system) and Tuflow SMS for a proposed removal of the pedestrian bridge downstream of Sunrise Boulevard over the American River in Sacramento, California. On April 4, 2013, from 5pm-6pm, four hydraulic project team members visited the site location and analyzed the existing conditions of the pedestrian bridge. It was 90 degrees Fahrenheit and sunny. Visibility was clear during that day.

II. Field Review

The team went to the site to observe the bed materials, vegetation, hydraulic issues, and channel irregularities. Geometric data for downstream pedestrian bridge was gathered. The data collected on the pedestrian bridge is seen on Table 1.

Table 1: Data Collected on Day of Study		
Collected Information	Value	Units
Deck Thickness	29	Inches
Span Length	30	Feet
Column Width	12.79	Inches
Number of Spans	11	NA
Abutment Slopes	45/45/90	Degrees

III. Design Requirements

Four simulations were required: low flow conditions with and without the pedestrian bridge, and high flow conditions with and without the pedestrian bridge. For the boundary conditions, the model estimates the flow of the river from about 200 feet downstream of the pedestrian bridge extending upstream of the Sunrise Bridge to the "Bridge Street" pedestrian bridge. Both high flow (90,000 cfs) and low flow (20,000 cfs) conditions will be analyzed. At the downstream location, the team assumed a starting water surface elevation of 85 feet for the high flow condition. For the low flow condition, the team assumed the flow is at normal depth at the downstream cross-section. A cross-section from SMS at the downstream limit was extracted and the team utilized Hydraulic Toolbox to determine the normal depth. The model was calibrated based on the flow rate and water surface elevation measured on the day of the April 4, 2013 field visit. The flow rate on the day of the field review was obtained from USGS. The water surface elevation at the low flow pedestrian bridge was measured during the site visit for calibration purposes.

A plan view of the location of study, is seen in Figure 1, was provided along with the survey scatter data in Figure 2. The photo depicts the boundary limits and topography of the area. The scatter data appeared over the topography when opened. Each scatter point denotes an elevation

IV. Location of Scatter Data and Scatter Point Data

Figure 1. Plan view of location of study

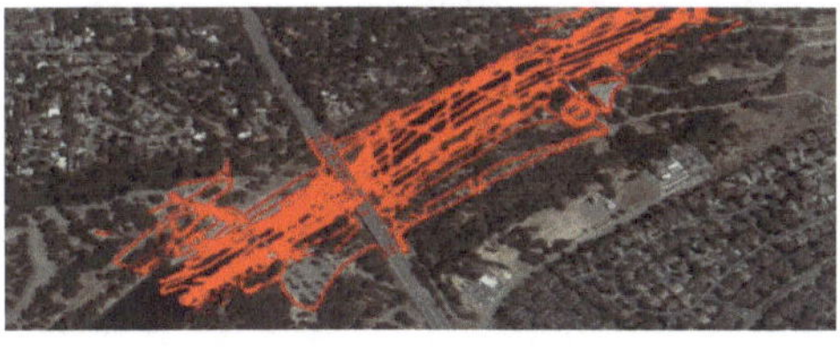

Figure 2. Scatter data provided

Project Details

A 2-dimensional grid containing 1 meter cells was created for the model. The upstream boundary was created north of the pedestrian bridge located on the east end of the map. The boundary type was set as flow versus time with a low steady flow of 566.34 m^3/s and a high steady flow of 2548.52 m^3/s. The downstream boundary was created south of the pedestrian bridge located on the west end of the map. The boundary type was set as water surface elevation versus time with a low elevation of 4.92 m and a high elevation of 25.91 m. The times for both low and high flow conditions at both boundaries were set to 24 hours. The 2-dimensional grid and the boundaries used may be seen in Figure 3.

5.1 Summary of Mannings N Values

The Manning's n-value for the rocks in the area of study was calculated using information derived with Hydraulic Toolbox 4.0. A rock/sediment gradation analysis was performed using a photo taken at the site. The photo used can be seen in Figure 4 to the right. The digital gradation analysis performed using Hydraulic Toolbox provided the D_{50} value to be used to calculate the Manning's n-value for the rocks shown in Figure 4. The D_5, D_{15}, D_{50}, D_{85}, and D_{100} values obtained from Hydraulic Toolbox may be seen in Figure 5.

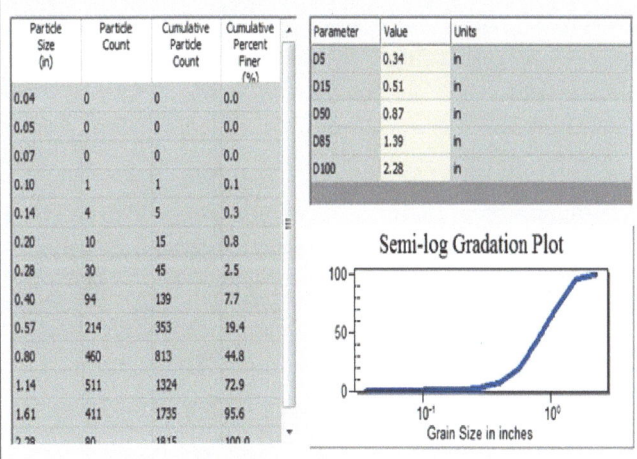

Particle Size (in)	Particle Count	Cumulative Particle Count	Cumulative Percent Finer (%)
0.04	0	0	0.0
0.05	0	0	0.0
0.07	0	0	0.0
0.10	1	1	0.1
0.14	4	5	0.3
0.20	10	15	0.8
0.28	30	45	2.5
0.40	94	139	7.7
0.57	214	353	19.4
0.80	460	813	44.8
1.14	511	1324	72.9
1.61	411	1735	95.6
2.28	80	1815	100.0

Parameter	Value	Units
D5	0.34	in
D15	0.51	in
D50	0.87	in
D85	1.39	in
D100	2.28	in

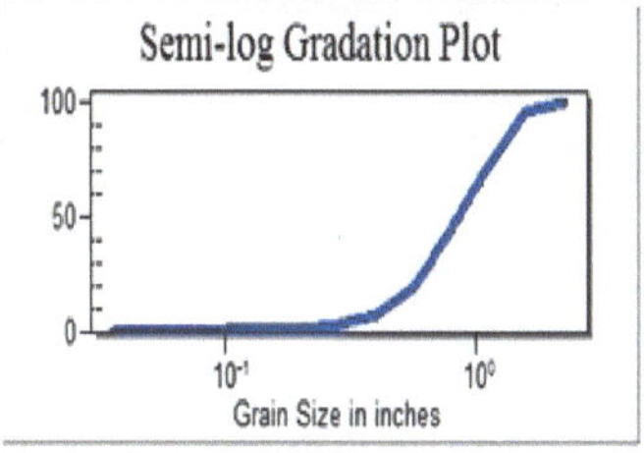

Figure 5: Results of Gradation Analysis

Using the D_{50} value obtained from Figure 5, the Manning's n-value was calculated using the following equation:

Where: n = Manning's n-value for stone.

D_{50} = Diameter of stone for which 50 percent by weight of the gradation is finer (ft).

Therefore the Manning's n-value for the stone in and around the river is the following:

Unknown Manning's n-values were obtained from tables. Material cover over the area of study was not created as a single material cover, but rather as separate covers. Creating a single material cover is unrealistic because more than one material may exist within a single area. For example, grass may exist underneath tree cover, or grass may grow in between rocks found underneath trees. In the cases mentioned, two to three materials exist within a single area. A separate n-value was assigned to each material. When it came time to run the Tuflow simulations, all material covers were dragged down and used as model components. The material covers with their respective Manning's n-values may be seen in Table 2.

Table 2: Material Covers and Manning's N-Values	
Material Cover	Manning's N-Value
Grassland	0.06
Gravel/Rocks	0.03
Trees/Brush	0.10
Concrete/Blacktop	0.02
River Channel	0.03

6.0 Run Results

Results of the 4 runs performed may be seen in Table 3. Included in the table are the parameters considered to be most important. The data presented was obtained from the observation cover created at the upstream face of the Sunrise Bridge.

Table 3: Important Parameters of Tuflow Simulations			
Run	Average WSEL (m)	Max Velocity (m/s)	Max Flow (m^3/s)
Low Flow without Ped Bridge	22.5	4.5	2520.0
Low Flow with Ped Bridge	23.0	4.5	3577.5
High Flow without Ped Bridge	26.5	7.0	7140.0
High Flow with Ped Bridge	28.0	4.0	4340.0

7.0 Longitudinal Profile Plots

The longitudinal profile plots were created using an observation cover that ran the length of the river, from boundary to boundary. The profiles for each run may be seen in Figure 6 through Figure 9. All plots are presented at a 1:15 time step from south to north with units in meters.

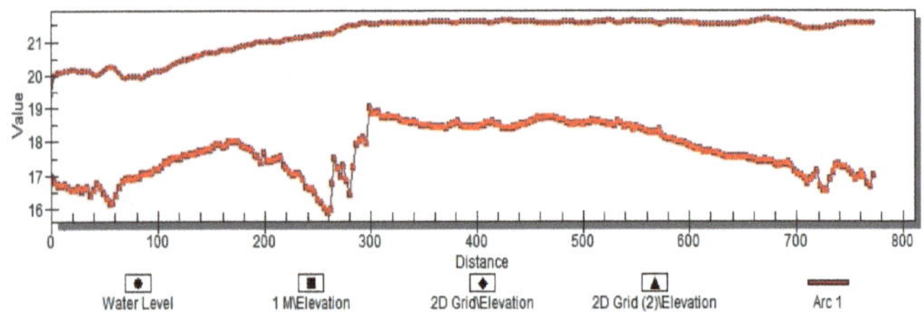

Figure 6: Profile Plot at Low Flow Conditions without the Pedestrian Bridge.

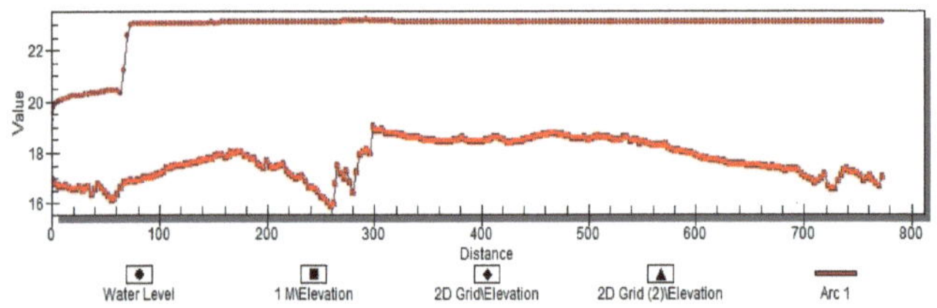

Figure 7: Profile Plot at Low Flow Conditions with the Pedestrian Bridge.

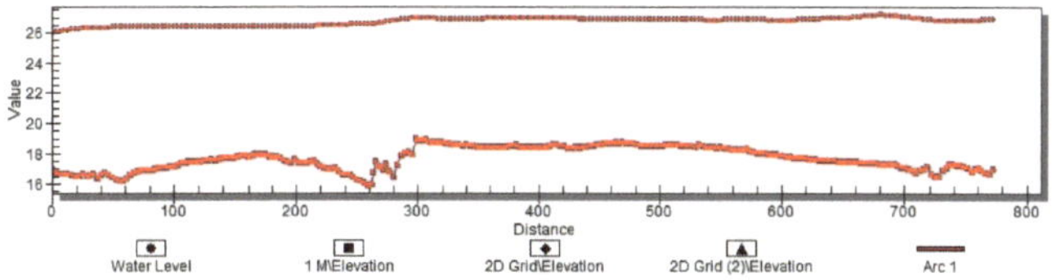

Figure 8: Profile Plot at High Flow Conditions without the Pedestrian Bridge.

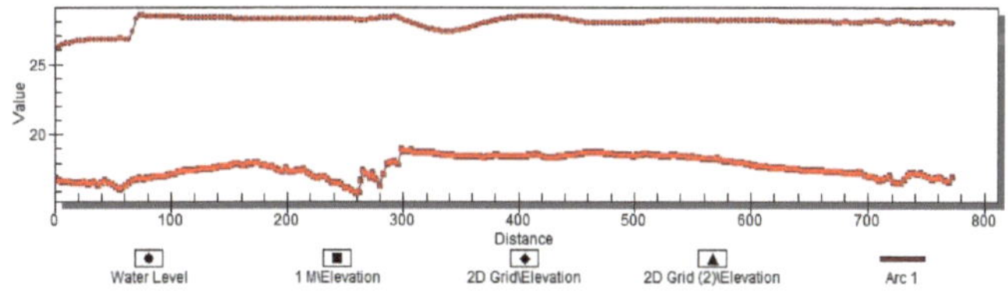

Figure 9: Profile Plot at High Flow Conditions with the Pedestrian Bridge.

8.0 Cross-Section Plots at Upstream Face of Sunrise Boulevard Bridge

The cross-section plots were created using an observation cover. All plots in Figure 10 through Figure 13 are presented at a 1:15 time step. The velocities highlighted in yellow are in meters per second. The elevations in meters are highlighted in red.

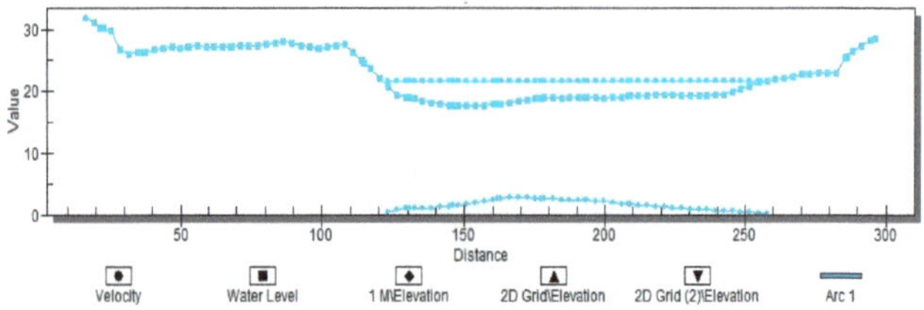

Figure 10: Velocity Cross-Section Plot at Low Flow Conditions without the Pedestrian Bridge

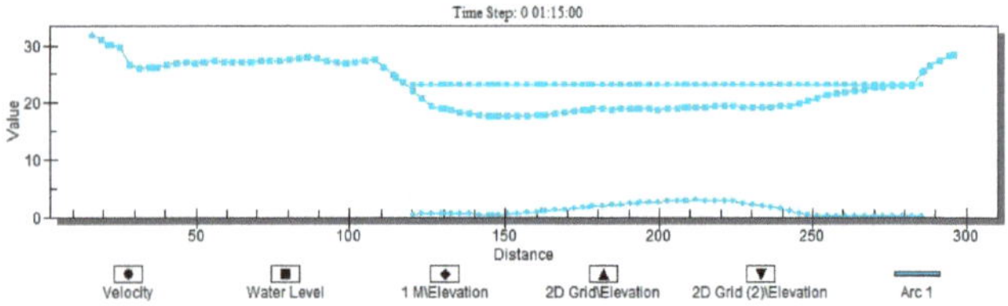

Figure 11: Velocity Cross-Section Plot at Low Flow Conditions with the Pedestrian Bridge

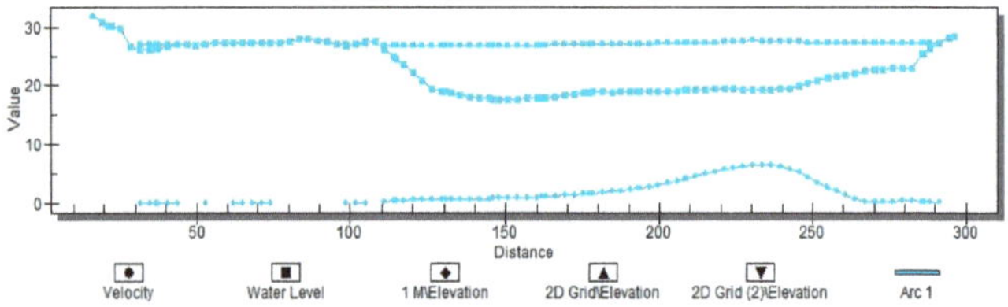

Figure 12: Velocity Cross-Section Plot at High Flow Conditions without the Pedestrian Bridge

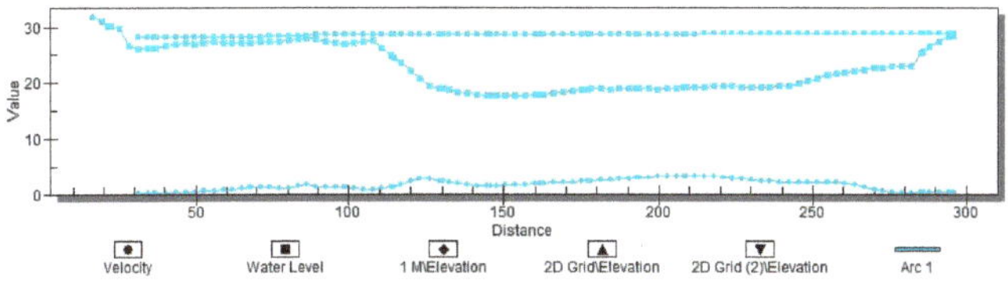

Figure 13: Velocity Cross-Section Plot at High Flow Conditions with the Pedestrian Bridge

Scour Calculations

The scour calculations were performed using the information presented in Table 4. Some K values were obtained using tables. K2 and K4 were calculated by hand. All relevant calculations are shown after Table 4.

Table 4: Values Used in Scour Calculations

Flow	90,000 cfs	Top Width	747 ft
Water Surface Elevation	45 ft	Flow Area	17938 ft^2
Max Velocity	8.4 ft/sec	Average Depth	24 ft
Average Velocity	7 ft/sec	Froude Number	$11.4/(32.2*45)^{.5}=0.30$

- $K_2 = (\cos\alpha + L/a \sin\alpha)^{0.65} = (\cos 24 + 30/13 \sin 24)^{0.65} = 1.4982$

- $v_{cD50} = K_u y^{1/6} * D_{50}^{1/3} = 11.17\ (45)^{1/6}(0.073)^{1/3} = 8.812$

- $v_{cD95} = K_u y^{1/6} * D_{95}^{1/3} = 11.17\ (45)^{1/6} * (0.19)^{1/3} =$

12.11

- $v_{icD50} = 0.645\ (D_{50}/a)^{0.053} * v_{cD50} = 0.645\ (0.073/4)^{0.053} * 8.812 = 4.60$
- $v_{icD95} = 0.645\ (D_{95}/a)^{0.053} * v_{cD95} = 0.645\ (0.19/4)^{0.053} * 12.11 = 6.65$
- $v_r = (v_{icD95} - v_{icD50}) / (v_{cD50} - v_{icD95}) = 6.65-4.60/ 8.812-6.65 = 0.9482$
- $k_4 = 0.4(v_r^{0.15}) = 0.4(0.9482)^{0.15} = 0.3968$
- $y_s = 2\ (a)K_1K_2K_3K_4(y/a)^{.35}Fr^{.43} = 2(4)(1)(1.4982)(1.1)(0.3968)(45/4)^{.35}(0.30)^{.43} =$ **7.32 ft**

4 DESIGN REPORT

Hydro Hornets
Water Treatment Team

DESIGN REPORT

Prepared For

2013 ASCE Mid-Pac Student Water Treatment Competition
San Jose State University
Attn: ASCE Daniel Wanner
One Washington Square
San Jose, CA 95192-0083

Prepared by

Sac State Water Treatment Team
California State University, Sacramento
6000 J. Street
Sacramento, CA 95819

March 15, 2013

CONTENTS

APPENDIX

1. INTRODUCTION

This design report details the development of the Sacramento State ASCE Water Treatment Club entry for the 2013 Mid-Pacific Student Water Treatment Competition. The Mid-Pacific Student Water Treatment Competition provides an opportunity for university students to develop skills in teamwork, design, and effective communication while exploring water-treatment practices. The team recognizes the growing need for young professionals with the capacity and experience beyond the classroom to contribute to the water and wastewater treatment fields. The Mid-Pac competition provided a framework for developing skills that will serve members in their professional development.

1.1 Competition Scenario

As the result of a wet year followed by a one-hundred year rainstorm, competing teams have been tasked with designing a wastewater-treatment system to treat household wastewater before dumping it into the Monterey Bay.

The water used will be an unspecified mixture of three different potable water purveyors. The team researched sources of water near the San Jose area. The group surmised that the large quantity of water necessary for competition would likely not be imported from a distant source. The team found that San Jose State has had source wells on campus since at least the 1940s. There are wells on the Main Campus and on the South Campus. San Jose Water Company serves as a backup to the campus wells, supplying water when one or more of the wells is off-line. The group also considered Santa Clara as a supplier due to its proximity to San Jose State. The consumer confidence and water-quality reports were studied to better understand the quality of the specific water used in the competition.

The influent used in this competition contains the following constituents:

- Pine Needles
- Dawn Dish Soap
- Gravel
- Tissue Paper
- Flour
- Potting Soil
- Lemon Juice
- Flush Wipes
- Tomato Juice
- Sand
- Yogurt
- Corn Starch

Resources available for construction of a chemical treatment and filtering system for wastewater in this scenario are limited to items found in a local hardware store. A complete list of materials can be found in the 2013 ASCE Mid-Pac Water Treatment Competition Rules (ASCE, 2011).

The goal of the competing designs is to produce the cleanest effluent with an emphasis placed on simplicity, sustainability, efficiency, structural and operational durability, creativity, and innovation. The water-quality parameters measured include pH, color, chlorine residual, nitrates, turbidity, dissolved oxygen, volume, and electrical conductivity.

1.2 Preparation for Mid-Pacific Competition

In preparation for the competition, a water treatment boot camp was organized to introduce fundamental water-quality concepts as well as to promote innovation through a hands-on wastewater treatment competition. The Water Treatment Boot Camp is an annual event that introduces environmental engineering concepts. A morning presentation by Richard Gutierrez and Jeff Riley from Carollo Engineers introduced real-life applications of water treatment. The presentation was followed by a lecture and lab administered by Dr. John Johnston, an Environmental Engineering professor at Sacramento State University. Dr. Johnston is also the Senior Technical Adviser for Sacramento State's Office of Water Programs Research Group, and his specialties include design, performance evaluation, and deployment of stormwater treatment technologies. His lecture and lab focused on key components of water quality such as pH, turbidity, and dissolved oxygen.

The afternoon portion was dedicated to the building of water-treatment systems. The group broke out into teams who built treatment systems and competed on the basis of creativity, pH, tubidity, and teamwork. The first hour was allotted for group brainstorming and planning, the second hour was allotted for building, and an additional half an hour was allowed for manipulation of pH with a variety of household items. Before loading the systems, design presentations were made to a panel of three judges from the professional community.

The event was a successful transition into preparation for Mid-Pac. Presentations and lectures were informative, labs were dynamic and illuminating, and attendance was ideal. Groups of 5 or 6 were large enough to practice teamwork while allowing individual hands-on experience. All attendees had the opportunity to not only design but use hand tools to build physical systems. In addition, attendees were able to practice presentation skills in front of professionals in a supportive context.

1.3 Strategy

Several members of Sacramento State's current design team competed in the 2012 Mid-Pacific Water Treatment competition. Knowledge gained through last year's experimentation and competition influenced this year's design. The effectiveness and ease of the stackable totes provided a working model for this year's design. Last year, extensive experimentation was performed attempting to mimic the effectiveness of Granular Activated Carbon (GAC) using charcoal. The team determined that the crushing of charcoal was overly time consuming and the resulting powder added color to the effluent. This year, the team was able to implement the use of GAC without time consuming experimentation.

Similar to last year, the primary concern for the filter system was clogging. To address this issue, a tote with a large surface area was placed on the top of the system for effective primary screening. The effectiveness of picking lime as both a flocculent and a modifier of pH were also discovered last year.

In order to maximize efficiency, the team was divided into several subteams focusing on physical system design, chemical treatment, sample tap design, documentation, research, design report, and presentation. This allowed team members to develop leadership and project management skills.

The team utilized faculty expertise and members from the professional community as resources for guidance and better understanding of water-quality concepts. The team would like to acknowledge Professor August Smarkel for clarifying principles and techniques in fluid mechanics, Dr. John Johnston for serving as faculty advisor, and Olivia Virdagamo (Engineer at Department of Water Resources) for advising the team on material selection and project management approach. Generous donations were provided by: Coleman Engineering, WaterWorks Engineers, GEI Consultants, Blankinship & Associates, Inc., and MBK Engineers.

2. FILTER DESIGN

The filter was designed as a four-tiered gravity-driven system, each tier addressing different water-quality parameters and competition targets. The design is modeled after Sac State's 2012 winning design with several additions and improvements. The primary focus of the design was maximizing particle removal, which was addressed using physical filtering processes in combination with chemical treatment. The secondary focus was water treatment, and the third focus was designing the sample tap. A schematic of the filter design is provided in Figure 2.1.

2.1 Physical Filtering Process

The physical filtering process was designed by first prioritizing materials and methods based on effectiveness, sustainability, and durability. Cost was an important concern, but not the primary focus due to its comparatively low point value in the competition. The four-tier system utilizes four, 18-gallon totes. Three of the 18-gallon totes are stackable and provide greater stability in comparison to traditional totes. The fourth 18-gallon tote is shallow but wide, creating a large surface area over which to place a quarter-inch wire mesh for prescreening. The increased surface area is an important improvement to the 2012 design because it effectively minimizes clogging concerns and provides a convenient place to suspend the sample tap.

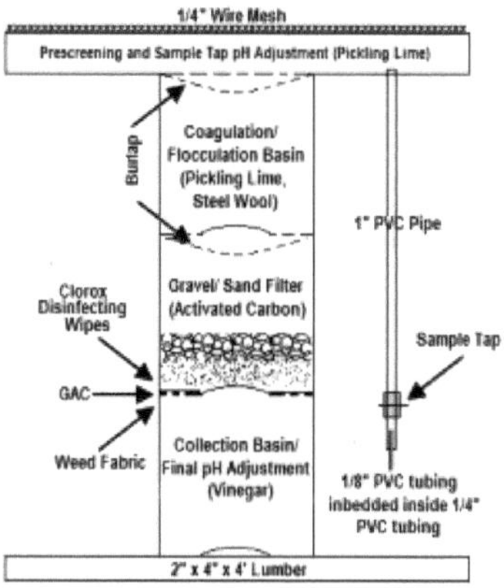

Figure 2.1 Schematic of 2013 Filter Design.

The holes in the bottom of the shallow top tote and sample tap opening are duct-taped in order to create a timing plug that is used in the 10-minute interaction time. When sample water is initially poured into the top tote, team members will safely lift the wire mesh and stir the water in the top tote for one minute (purpose of stirring is discussed in the *Chemical Treatment* subsection of this report). After 1 minute, the duct tape for the sample tap is pulled in order to fill the attached 1-inch PVC pipe with just over 500 mL of sample water (tap is further discussed in the *Sample Tap Design* subsection of this report). After the PVC pipe for the tap is filled (indicated when bubbles cease to emerge on the water surface directly above the PVC pipe), the duct tape covering the holes drilled at the bottom of the top tote are pulled.

When the timing plug (duct tape) for the top-tote holes is pulled, sample water is filtered through coarse burlap to capture large particles before it enters the second tote. The second tote also utilizes a timing plug. In the second tote, team members will rapidly mix the water to kick start the coagulation process with pickling lime and lowering of dissolved oxygen with steel wool for one minute followed by a slow mix of four to six minutes. After slow mixing, the timing plug is pulled and water is released into the third tote.

The third tote starts off with another layer of burlap suspended from the second tote to capture large particles before going through the gravel and sand filter. Weed fabric is duct-taped over perforations in the bottom of the tote to hold back sand and large particles from entering the collection basin. Directly on top of the weed fabric, a thin layer of GAC is placed to aid in removing color from water. Another layer of weed fabric separates the GAC from the sand. Gravel is placed on top of the sand. Gravel and sand are effective filters for catching floccs developed in the second tote mixing. Water filters through the gravel, sand, GAC, and weed fabric into the collection basin.

The fourth tote collects the filtered water. Vinegar is added at the bottom of this tote to provide final adjustment of the pH, which was raised during the coagulation/ flocculation process in the second tote.

2.2 Chemical Treatment

In order to reach the competition target values for the various water-quality parameters, several chemical treatments were incorporated into the treatment process design. The water-quality parameters evaluated in this design report include pH, color, chlorine residual, nitrates, turbidity, dissolved oxygen, electrical conductivity, and returned volume. The primary strategy for chemical treatment was to focus on particular water-quality parameters that had the easiest targets to achieve. All water-quality parameters were considered in the balancing act of chemical treatment, but the team focused on turbidity, pH, dissolved oxygen, color, and nitrates. The team discovered that improvements in one area often negatively effected other water-quality parameters.

2.2.1 Turbidity and pH

Turbidity was the easiest water-quality parameter to address. Pickling lime (calicium hydroxide) served dual purposes: as a coagulant to form floccs that can be more readily removed by physical filtering processes such as burlap and sand, and as a pH modifier. Pickling lime releases positive ions of calcium into the water which helps negatively charged pollutant particles to neutralize their surface ions and clump together.

The team also evaluated other coagulants and oxidants. The team observed some positive effects of polymeric materials present in toothpaste for coagulation, but due to its slow effect, the team decided not to use toothpaste for this purpose. Hydrogen peroxide was explored in order to oxidize a few pollutants and encourage faster flocculation, but hydrogen peroxide increases dissolved oxygen in excess of the target value.

In order to determine the amounts of pickling lime to balance effectiveness and cost as well as modify pH, varying amounts of pickling lime was added to six 2-L samples of water with 1 minute of fast mix and 10 minutes of slow mix. After mixing, each sample was poured through a sand filter and tested for pH and turbidity; results are shown in Table 1.

Table 1. Turbidity and pH of varying concentrations of pickling lime.

Concentration (g/2L)	0.25	0.5	1.0	1.5	2.0	2.5
Turbidity (NTU)	1000+	633	580	340	325	323
pH	5.58	8.09	10.54	11.58	11.89	11.98

The results in Table 1 provided ball-park values for testing the system. The final concentrations of pickling lime used in the treatment system is 13 grams per 9 gallons of influent in the top tote (pH = 7.27), and 21 grams per 9 gallons in the second tote.

2.2.2 Dissolved Oxygen

Traditionally, water quality is improved by increasing dissolved oxygen (DO); for this competition, however, the target value was 5 mg/L which prompted investigation into materials that lower DO. Although bacteria from the yogurt in the influent lowers DO due to the biological oxygen demand (BOD), water trickling through the filter has increased surface area contact with air and therefore increases the DO. The team discovered that steel wool was effective at lowering DO because of rusting. Rusting occurs due to the bonding between oxygen atoms present in the water with ion atoms. This phenomenon helped the team effectively reduce the DO level in the system towards the target value.

2.2.3 Color

Of the water-quality parameters focused on, color was the most difficult target value (15 color units) to achieve but the team focused on lowering it as much as possible by using GAC. GAC was placed underneath the sand filter and spread out to increase surface area and contact time with the water. GAC is known to efficiently adsorb color bodies in water (Joyce, 1966). While research shows both turbitity and color are positively affected by a deeper sand bed (Muhammed, 1996), it also negatively affects filtration time, DO, chlorine residual, and return volume.

2.2.4 Nitrates

To reduce nitrates, the water filters through several layers of Clorox Disinfectant wipes. The two active ingredients in these wipes are dimethyl benzyl ammonium chloride and diethyl benzyl ammonium chloride.

When the water filters through, the nitrates react with the ammonium chloride in the following unbalanced reaction:

$$NO_3^- \text{ (aq)} + NH_4Cl \text{ (aq)} \dashrightarrow N_2 \text{ (g)} + H_2O \text{ (l)} + Cl^- \text{ (aq)}$$

In this double displacement reaction, the nitrates react with the ammonium to form ammonium nitrate which reduces to nitrogen gas and water when slightly heated. The more ammonium, the greater the reduction in nitrates. No heat could be added to the water treatment system so reaction is constrained.

2.2.5 Other Water-Quality Parameters

No chlorine was observed in initial testing of influent using total chlorine reagent, but some chlorine was added with the addition of Clorox Disinfectant Wipes used to treat for nitrates. Electrical conductivity was the most difficult water-quality parameter to address. Chemical treatment that improved one parameter usually increased electrical conductivity. In order to maximize the effectiveness of chemical treatment for the other water-quality parameters, electrical conductivity was largely ignored.

3. SAMPLE TAP DESIGN

During the first 10 minutes of the loading phase, 500 mL of water will be collected from the sample tap. The sample tap effluent was designed to have an average flow rate of 1 liter per minute (500 mL per 30 seconds) and a pH of 7. Since the pressure head in the system is variable over time and unpredictable, a fixed volume of water was collected during the first few minutes of the loading phase to be used for the tap. A schematic of the sample tap is shown in Figure 3.1.

The sample tap consisted of descending sizes of several pieces of PVC pipe and tubing. A 1-inch PVC pipe was cut to 31 inches with one end "screwed" into the top tote used during the primary screening and the other end was attached to a PVC cap with a small hole drilled on the bottom. One end of the ¼-inch PVC tubing was inserted into the hole drilled into the PVC cap, the other end was fastened to a ¼-inch PVC ball valve. At the other end of the PVC ball valve, 1/8-inch PVC tubing, embedded within 1/4-inch tubing, was attached to produce the required flow rate. The sample tap consistently produced 500 mL of sample water in 30 seconds.

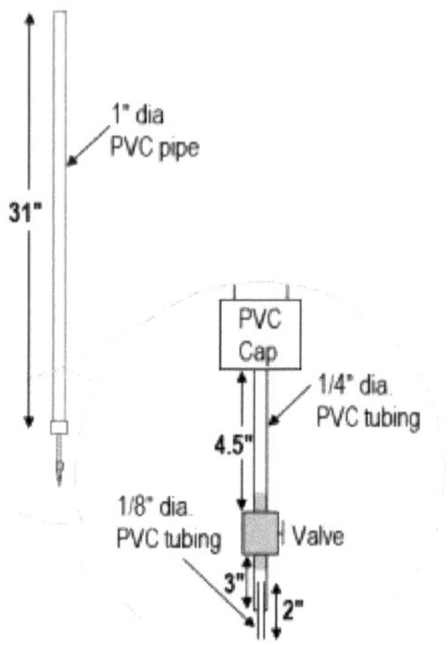

Figure 3.1 Schematic of Sample Tap Design.

The team also attempted to calculate flow rates using the energy equation (Eqn. 1). The calculations included major and minor head losses induced by the constrictions and friction losses caused by the wall of the pipes. This provided the opportunity to explore fluids concepts and was an introduction to resources like the moody diagram. The calculated flow rate differed from the actual flow rate in excess of 50% and was therefore excluded from the report. The team surmised that some of the error resulted from assumptions such as the smooth wall of the pipe and variation caused by ragged edges of the pipe. The team benefited from the academic exploration but chose to proceed with trial and error.

$$z_1 + \frac{P_1}{\gamma} + \frac{v_1^2}{2g} = z_2 + \frac{P_2}{\gamma} + \frac{v_2^2}{2g} + h_{L(friction)} + h_{L(constriction)} \qquad \text{(Eqn. 1)}$$

4. MATERIALS AND COST ANALYSIS

A list of materials and associated costs are included in Appendix A1. Materials were chosen based on reuse from last year, effectiveness, durability, sustainability, and cost. Evaluations of chosen materials were discussed in the filter design section of this design report. The total cost for the water treatment system including materials, labor, measuring equipment, and safety equipment is $175.26.

5. SUSTAINABILITY

The team utilized several sustainable materials and practices. Reusing materials from previous years was the easiest and most cost-effective sustainable practice. The stackable totes are post-consumer recycling number 5 (PP5) which refers to polypropylene plastics. PP5 materials can be recycled into items such as signal lights, brooms, ice scrapers, bicycle racks, bins, and rakes (Howard, 2011). The pickling lime used in the system contains 30 percent calcium hydroxide which is non toxic at the concentrations used in the system. This is important because other possible products used as a coagulant have more detrimental effects on the environment and aquatic life. With the exception of duct tape, all of the materials were chosen based on their durability and can be reused over long periods of time.

6. ENVIRONMENTAL IMPACTS

The quality of water discharged into the environment from communities and industries has a significant impact on aquatic ecosystems and public health . During system design the team tried to balance target water-quality parameters with ecologically responsible system design. Through research, the team became aware of the listing of the steelhead trout and Coho salmon as federally threatened species. They are a valued part of the habitat of the San Lorenzo River. Water containing a pH level higher than 7-8 can cause damage to the skin and eyes of aquatic life. Imbalanced pH also has the potential for fish fatality.

Higher pH levels can compound the effects of several other toxic substances. For instance, ammonia is 10 times more severe at pH 8 than at pH 7. In order to maintain a balanced pH level the team used vinegar to neutralize the pH.

Although fish and marine life thrive on a high level of dissolved oxygen in the water, the target level for the competition was lower than what was measured in the influent. To reduce the DO in the water steel wool was used. The oxidation induced by the rusting process effectively lowered the DO in the water.

7. SUMMARY AND CONCLUSIONS

Through research, consultation, experimentation, and system design, the team devised a simple, constructible, cost-effective wastewater treatment system. Participation in the Mid-Pacific Student Water Treatment Competition included attending a wastewater Boot Camp, chemical experimentation, as well as collaborative system design. These processes all contributed to the development of a valuable skill set that will serve members of the Sacramento State ASCE Water Treatment Club in their future careers in the water industry.

8. REFERENCES

American Society of Civil Engineers, 2012, Water Treatment Competition Rules: ASCE Mid-Pac Mid-Pacific Student Conference, San Jose State University and Santa Clara University, December 15, 2012.

City of Santa Cruz (2003). San Lorenzo Urban River Plan: A Plan for the San Lorenzo River, Branciforte Creek and Jessie Street Marsh: City of Santa Cruz San Lorenzo Urban River Plan Task Force, Santa Cruz, California, June 24, 2003.

Competition Rules; ASCE Mid-Pac Conference; San Jose State University, CA; Santa Clara University, CA.

Howard, B.C., (2011). What Do Recycling Symbols on Plastics Mean?: The Daily Green website. Available at *http://www.thedailygreen.com/green-homes/latest/recycling-symbols-plastics-460321#slide-1*

Joyce, R. S.; Sukeni, V. A. , "Feasibility of Granular Activated Carbon Adsorption for Waste Water Renovation", publ. no.999WP-12, US public health service, Washington D.C. (1964).

Muhammad, M.; Ellis, K.; Parr, J.; Smith, M.D (1996). Optimization of slow sand filtration. Reaching the unreached: challenges for the 21st century. 22nd WEDC Conference New Delhi, India, 1996. pp.283-5. Available at *http://wedc.lboro.ac.uk/resources/conference/22/Muhamme.pdf*

San Jose State University. Campus Water Resources: Facilities Development and Operations, San Jose State University, San Jose, California. Available at *http://www.sjsu.edu/fdo/energy/sustainability/water/*

San Jose State Unversity (2011). Consumer Confidence Report: San Jose State University, San Jose, California. Available at *http://www.sjsu.edu/fdo/docs/sjsu_main_campus_ccr.pdf*

San Jose Water Company (2011). Investing in Water Quality Annual Water Quality Report: San Jose Water Company, San Jose, California, 2011. Available at *http://www.sjwater.com/files/documents/WaterQualityReport_2011.pdf*

Hydraulic
SOLUTIONS

Hydraulic Solutions
6000 J Street
Sacramento, CA 95819-6029

April 5, 2013

California Department of Water Resources
Division of Flood Management
Flood Maintenance Office
3310 El Camino Avenue, Room 110
Sacramento, CA 95821

Attention: Julius D. Bautista, P.E.

Hydraulic Solutions has prepared the following technical report regarding the hydraulic analysis procedure and development of alternatives for the Sacramento Weir and Bypass improvements. The hydraulic analysis, as required by the United States Army Corps of Engineers (USACE), West Sacramento Area Flood Control Agency, the City of West Sacramento, and California Department of Water Resources includes the analysis of the 100-year and 200-year storm event water surface elevations and levee freeboard for the existing conditions, a 900 foot bypass expansion, and an 1800 foot bypass expansion and respective extensions of the weir. Additionally, Hydraulic solutions conducted a review of possible weir gate alternatives, which included inflatable weir gate systems and radial weir gate systems, and providing a curved bypass connection to the Yolo Bypass for the 900 foot bypass expansion alternative. Each team member of Hydraulics Solutions worked together collaboratively to assemble the Sacramento Weir and Bypass Improvements: Hydraulic Analysis Report. Eduard Egov and Patrick Ervin worked together as the hydraulic modeling team to run the USACE Sacramento River Basin HEC-RAS models, and produce the water surface profile and levee freeboard plots and reviewed the curved bypass alternative. Matt Todd, Grace Lau, Martin Lee, and Hani Nour worked together as the hydraulic writing team which included researching and reviewing the weir gate alternatives, operation and maintenance requirements of the existing levees, weir, and bypass, and report writing. Keith Jones managed both work groups.

Sincerely,

Keith Jones
Matt Todd
Grace Lau
Patrick Ervin
Martin Lee
Hani Nour
Eduard Egov

Enclosure: Sacramento Weir and Bypass Improvements: Hydraulic Analysis Report

Sacramento Weir and Bypass Improvements: Hydraulic Analysis Report

Presented To:

United States Army Corps of Engineers

West Sacramento Area Flood Control Agency

The City of West Sacramento

California Department of Water Resources
Division of Flood Management
Flood Maintenance Office
Julius Bautista, P.E.

Presented By:

Hydraulic Solutions
6000 J Street, Sacramento, CA 95819-6029
(925) 759-9850
email@HydraulicSolutions.com

TABLE OF CONTENTS

LIST OF FIGURES

LIST OF TABLES

ACRONYMS AND ABBREVIATIONS

1957 Design Profile	Levee and Channel Profile File Number 50-10-3334
cfs	Cubic Feet per Second
Comp Study	Sacramento and San Joaquin River Basins Comprehensive Study
CVFPB	Central Valley Flood Protection Board (Formerly the Reclamation Board)
CVFPP	Central Valley Flood Protection Plan
DWR	California Department of Water Resources
DFM	California Department of Water Resources Division of Flood Management
FOC	California Department of Water Resources Division of Flood Management Flood Operations Center
ft	Feet, Foot
GIS	Geographic Information System
HEC-RAS	Hydrologic Engineering Center - River Analysis System (United States Army Corps of Engineers)
HEC-HMS	Hydrologic Engineering Center - Hydrologic Modeling System (United States Army Corps of Engineers)
LiDAR	Light Detection and Ranging
O&M Manual	Supplement to Standard Operation and Maintenance Manual Sacramento River Flood Control Project, Unit No. 116 Sacramento Bypass South Levee, Unit No. 122 Part 1 Sacramento Bypass North Levee, Unit No. 158 Sacramento Weir
"n" values	Manning's Roughness Coefficient Values
NAD 83	North American Datum of 1983
NAVD 88	North American Vertical Datum of 1988
NGVD 29	National Geodetic Vertical Datum of 1929
RM	River Mile
SMY	California Department of Water Resources Sacramento Maintenance Yard
SRFCP	Sacramento River Flood Control Project

SPFC	State Plan of Flood Control
stage	Water level as read by gage
TOL	Top of a Levee
USACE	United States Army Corps of Engineers
USED	United States Engineering Datum
USGS	United States Geological Survey
WSAFCA	West Sacramento Area Flood Control Agency
WSEL	Water Surface Elevation

EXECUTIVE SUMMARY

At the request of the California Department of Water Resources (DWR), Hydraulic Solutions prepared the following hydraulic analysis report for the proposed Sacramento Weir and Bypass project. The Sacramento Weir and Bypass project was one part of the 2012 Central Valley Flood Protection Plan (CVFPP). The CVFPP will bring the Sacramento Region into compliance with California Senate Bill 5, which required all municipalities to create a plan to provide flood protection for a 200-year storm event by 2015 and to be in effect by 2025. The Sacramento Weir and Bypass project will protect the Sacramento, West Sacramento, and the Natomas regions from floodwaters escaping the Sacramento River by increasing the volume of flow diverted through the weir and bypass to maintain acceptable river stages and levee freeboard in the larger storm event. Hydraulic Solutions performed a hydraulic analysis, using HEC-RAS modeling, of the 100-year and 200-year storm event water surface elevations and levee freeboard for the existing conditions, a 900 foot bypass expansion, and an 1800 foot bypass expansion and respective extensions of the weir. Additionally, Hydraulic Solutions conducted a review of possible weir gate alternatives, which included inflatable weir gate systems and radial weir gate systems, and any potential effects of providing a curved bypass connection to the Yolo Bypass for the 900 foot bypass expansion alternative. The proposal of the curved alignment for the Sacramento Bypass connection into the Yolo Bypass was suggested to avoid conflicts with the existing landfill to the north. Review if the weir gate alternatives were recommended to identify an automated weir gate option which reduced the health and safety risk to DWR maintenance personnel and increased the efficiency of the gate operation.

Based on the hydraulic analysis, Hydraulic Solutions recommends the 1800 foot bypass expansion and weir extension alternative. The 1800 foot bypass expansion will reduce the water surface elevations within the Sacramento Bypass and the Sacramento River more than the reduction associated with the 900 feet expansion alternative, which would help meet the 6 foot and 3 foot freeboard requirements, respectively. The increase in free board will reduce the required levee heights and associated levee improvements. The levee heights within the Sacramento bypass channel were found to be deficient by 5 feet to meet the required 6 foot freeboard and Hydraulic Solutions recommends levee improvements to increase the levee height to meet the freeboard requirements. The increase in levee height may affect the California Highway Patrol property to the south and the agricultural properties to the north, requiring the need to obtain additional right of way or record a levee or slope easement. Hydraulic Solutions also recommends using an automated inflatable weir gate system to replace the existing manually operated wooden needle gates. The inflatable weir gate will allow an operator in an office to monitor the gate operation, relieving the health and safety concerns related to the current weir gates. The inflatable weir gates require minimal structural support, which will keep structural cost low.

In conclusion, Hydraulic Solutions has recommended that the Sacramento Weir and Bypass project consist of expanding the bypass and extending the weir by 1800 feet, removing and replacing the weir gates with automated inflatable weir gates and installing these gates on the extended sections of the weir, and to increase the north and south levee heights within the Sacramento Bypass by 5 feet to meet the required 6 foot levee freeboard requirement.

1.0. INTRODUCTION

In 2007, Governor Arnold Schwarzenegger signed a six-bill package that identified high-risk flood hazard zones and planned for future development. On June 29, 2012, the Central Valley Flood Protection Board (CVFPB), formally the State Reclamation Board, adopted the Central Valley Flood Protection Plan (CVFPP) for the Sacramento-San Joaquin River Flood Management System, as required by California State Senate Bill SB5 and the Central Valley Flood Protection Act of 2008. The CVFPP proposes a system wide approach for integrated flood management in areas currently protected by facilities of the State Plan of Flood Control (SPFC), including weir, bypass, and levee structures to be designed for a 200-year storm event. The California Department of Water Resources (DWR), the United States Army Corp of Engineers (USACE), the West Sacramento Area Flood Control Agency (WSAFCA), and the City of West Sacramento have combined resources to evaluate the rehabilitation of the Sacramento Weir and Bypass.

1.1. BACKGROUND

Built in 1916, the Sacramento Weir is one of ten overflow structures along the Sacramento River used to prevent flooding of adjacent populated areas, more specifically, the City of Sacramento. The Sacramento Weir is manually operated with 48 gates which can be opened as required to prevent the stage of the Sacramento River at the I Street gage from exceeding 29 feet or to hold the stage at the downstream end of the Sacramento Weir to 27.5 feet during storm events. The weir gates are opened when the Sacramento River stage rises to 27.5 feet National Geodetic Vertical Datum of 1929 (NGVD 29) at the I Street Bridge, with the forecast showing a continued increase in precipitation. When the weir gates are opened, floodwaters from the Sacramento River are diverted into the Sacramento Bypass which directs the flow into the Yolo Bypass, which then reconnects with the Sacramento River in Rio Vista.

1.2. NEED AND PURPOSE

By approximately 2015, the CVFPP requires all cities and counties within the Sacramento-San Joaquin Valley to make a finding regarding an urban level of flood protection when considering decisions about entering into a development agreement for a property; approving a discretionary permit or entitlement for any property development or use, or approving a ministerial permit that would result in construction of a new residence; or approving a tentative map/parcel map for a subdivision. This urban level of flood protection means the level of protection that is necessary to withstand a 200-year flood event within any given year. (Urban Level of Flood Protection Criteria. 2012)

The Sacramento Weir and Bypass project is one part of the 2012 CVFPP which will help to meet criteria for attaining an urban level of flood protection. The Sacramento Weir and Bypass project will protect the cities of Sacramento and West Sacramento, as well as the Natomas region from floodwaters escaping the Sacramento River by increasing the flows diverted through the weir and bypass to maintain acceptable river stages. Additionally, the weir's manual operation is cumbersome and dangerous to the operators due to the timing of the opening and closing of the weir gates. The Sacramento Weir project will automate the weir gates to eliminate the danger of the manual operation and increase the efficiency of opening and closing the gates. Hydraulic Solutions will review several design alternatives and provide a recommendation for the improvement of the aging Sacramento Weir, including an automated weir gate, and expansion of the Bypass. (Zenobia, Kent. 2012)

1.3. LOCATION

The Sacramento Weir and Bypass are located in Yolo County, California along the right bank of the Sacramento River near Bryte, California. The weir and bypass are located approximately 4 miles upstream of the Tower Bridge, and about 2 miles upstream from the mouth of the American River. Figure 1.1 is a vicinity map of the Sacramento Weir and Bypass.

1.4. EXISTING OPERATIONS AND MAINTENANCE

The Sacramento Weir consists of 48 weir bays mounted with wooden needle weir gates. Each weir gate contains 38 wooden needles that are bottom hinged on the downstream side and held into a closed position with a metal holding beam that is latched to the weir wall. Figure 1.2 is a picture of the existing weir structure showing the wooden needle gate in the closed position. DWR's Sacramento Maintenance Yard (SMY) manually opens the gates by physically activating the latch that releases the metal holding beam. The beam is connected to a wire attached to the weir structure to prevent the beam from being carried away in the flow. SMY staff retrieves the beam and resets the wooden needles to close the gate. This process of closing the existing weir gates is extremely difficult and dangerous to perform because it takes multiple workers and a crane to place the gate back into its closed position while water continues to flow through the weir. The gates are closed as quickly as is practicable once the river stage falls below 25.0 feet NGVD 29. This allows a flushing flow to re-suspend sediment deposited within the Sacramento River between the Sacramento Weir and the American River.

The timing and number of gates to be opened or closed is communicated to the SMY through the DWR's Division of Flood Management (DFM) Flood Operations Center (FOC), which performs river stage and precipitation forecasting. The weir gates are opened and closed to meet one of two criteria: (1) to prevent the stage at the I Street gage from exceeding 29 feet, or (2) to hold the stage at the downstream end of the weir to 27.5 feet. These criteria have been determined to balance two goals: (1) minimize sediment deposition due to decreased flow velocities downstream from the weir to the mouth of American River; and (2) limit the flooding of agricultural lands in the Yolo Bypass until after they have been inundated by floodwaters over Fremont Weir. Though the weir crest elevation is 24.75 feet, the weir gates are not opened until the river reaches 27.5 feet at the I Street gage with a forecast showing continual rise of the river stage. The I Street gage is located about 1,000 feet upstream from the I Street Bridge and about 3,500 feet downstream from the mouth of the American River. Once all 48 gates are open, Sacramento River stages from Verona to Freeport may continue to rise during a major storm event. Project design stages are 41.3 feet at Verona, 31.5 feet at the south end of the Sacramento Weir, and 31 feet at the I Street gage. All river stages are identified with respect to NGVD 29. (Sacramento River Flood Control Project Weirs and Flood Relief Structures. 2010)

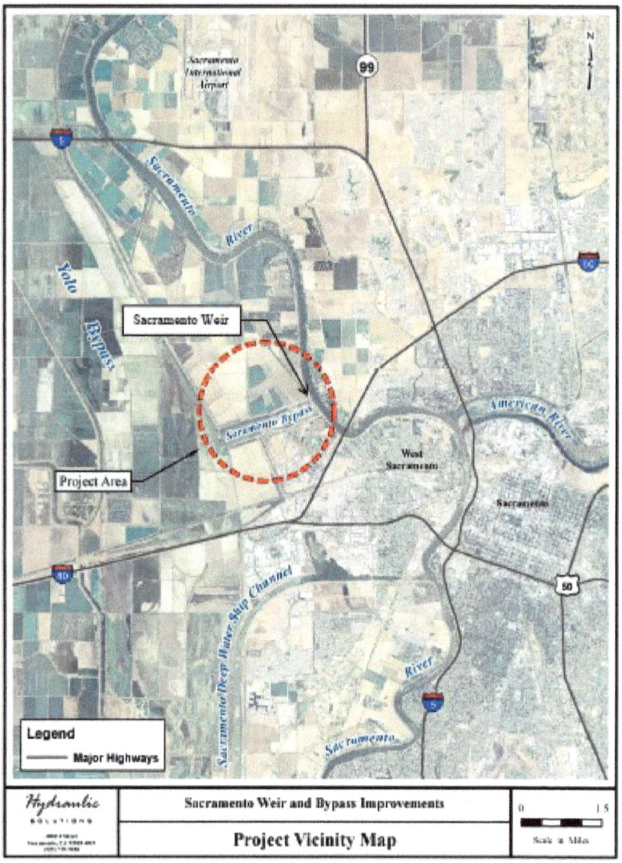

99
Sacramento River
Yolo Bypass
Sacramento Weir
Project Area
Sacramento Bypass
American River
West Sacramento
Sacramento
80
50
Sacramento Deep Water Ship Channel
River
Sacramento
Legend
Major Highways
Hydraulic SOLUTIONS
Sacramento Weir and Bypass Improvements
Project Vicinity Map

Figure 1.1

Figure 1.2: Sacramento Weir Gates (Closed Position)

The USACE and the DWR, pursuant to California Water Code Section 8361, are responsible for the operation and maintenance of the flood management improvements constructed with the Sacramento Flood Control Project. Under Sections 8361(d) (e) (f) (j), DWR shall operate and maintain the Sacramento Weir and the Sacramento Bypass Channel, including the levees. Project specific operation and maintenance requirements are found in the USACE Supplement to Standard Operation and Maintenance Manual.

Maintenance of the Sacramento Weir and Sacramento Bypass, including levees, is the responsibility of the SMY. Maintenance responsibilities include, but are not limited to the following:

- Visual inspection of the weir, bypass, and levee structures for scour, sediment deposition, vegetation, and structure conditions.
- Remove sediment deposits from stilling basin, as needed.
- Repair or replace weir parts, as needed.
- Repair levee, as needed.
- Remove vegetation, as needed.

Typically, maintenance will occur outside the flood season to allow full inspection of the bypass and structures when they are not in use and to limit health and safety hazards to SMY personnel.

1.5. DESIGN CRITERIA

The CVFPP requires all cities and counties within the Sacramento-San Joaquin Valley to attain urban level flood protection for a 200-year flood event. The current means of flood protection for the West Sacramento region includes a system of levees along the east and west banks of the Sacramento River. The Sacramento Weir and Bypass acts as a relief structure to reduce the river stage to help prevent the overtopping of the levees within the Sacramento River.

The hydraulic review of the Sacramento Weir and Bypass project required modification to the existing Hydrologic Engineering Center - River Analysis System (HEC-RAS) model to account for the proposed 900 and 1,800 foot bypass expansion alternatives of the Sacramento Bypass for the current standard of a 100-year and 200-year flood event, which is required by the CVFPP. The review will include an analysis of the resulting water surface elevations to determine if proposed improvements meet levee freeboard requirements. The required levee freeboard is 3 feet within the Sacramento and American Rivers. The required levee freeboard is 6 feet within the Sacramento Bypass.

The hydraulic review also considered the automation of the weir gates for the Sacramento Weir. Gate closure can take up to an hour per gate, affecting the efficiency of the weir. The alternative gates considered were fully automated; physical interaction with the gate structure is not required to open and close the gates. The gates were required to open and close quickly and efficiently, provide similar detention abilities to the existing gates, and have simpler maintenance requirements.

1.6. ALTERNATIVES

The alternatives considered and reviewed included expansion of the existing bypass channel, the removal and extension of the existing weir structure, and replacement of the existing weir gates. The alternatives to the bypass channel include 900 and 1,800 foot expansions, achieved by offsetting the north levee, and a 900 foot expansion with a curved modification to the Yolo Bypass. The removal and replacement of the weir structure includes the proposed 900 and 1,800 foot expansions. The gate system review includes two automated systems, which are the inflatable weir gate system and the radial weir gate system.

1.6.1. WEIR STRUCTURES

The existing weir gates are a manually operated system that is time consuming and hazardous for maintenance personnel to open and close. Two alternatives which provide an automated system were reviewed.

1.6.1.1. INFLATABLE WEIR GATE SYSTEMS

Inflatable weir gate systems consist of an inflatable rubber bladder which supports a row of steel gate panels. The gates open and close by controlling the air pressure within the bladders. The gates can be finitely adjusted within a user specified control range to maintain desired water surface elevations. See Figure 1.3 for a typical cross section of an inflatable weir gate.

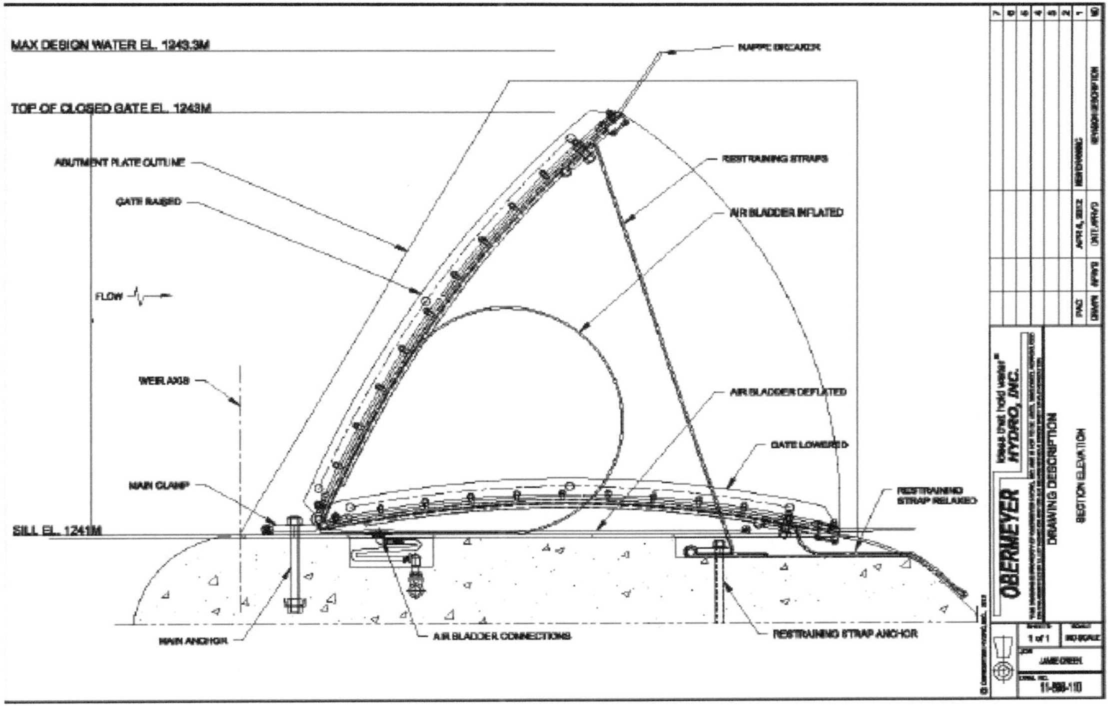
MAX DESIGN WATER EL. 1243.3M
TOP OF CLOSED GATE EL. 1243M
ABUTMENT PLATE OUTLINE
GATE RAISED
FLOW
WEIR AXIS
MAIN CLAMP
SILL EL. 1241M
MAIN ANCHOR
AIR BLADDER CONNECTIONS
RESTRAINING STRAP ANCHOR
RESTRAINING STRAP RELAXED
GATE LOWERED
AIR BLADDER DEFLATED
AIR BLADDER INFLATED
RESTRAINING STRAPS
NAPPE BREAKER
OBERMEYER
Ideas that hold water
HYDRO, INC.
DRAWING DESCRIPTION
SECTION ELEVATION
1 of 1
NO SCALE
JAKE GREEN
11-089-110
PAO
APR 4, 2012
DRAWN
APRV'D
DATE APRV'D
REVISION DESCRIPTION
NO

1.6.1.2. RADIAL WEIR GATE SYSTEMS

Radial gates are designed for wide and unobstructed waterways. These gates are light in weight and require minimal hoist effort to open and close. Radial gates can be used for upstream level control in open channel installations. Radial gates are designed such that the flow is always one-way, against the face of the gate. See Figure 1.4 for a typical cross section of an inflatable weir gate.

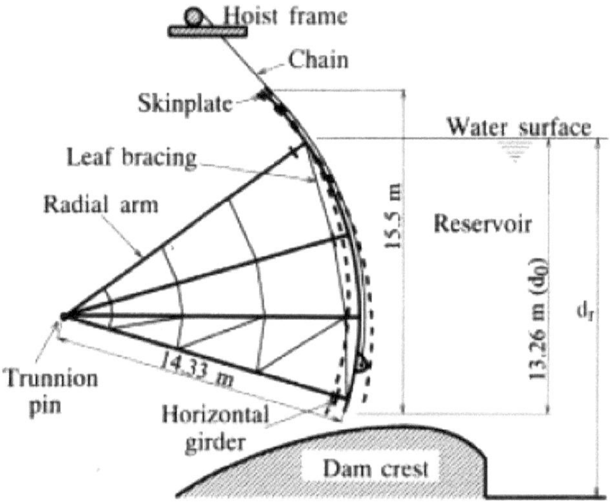

Figure 1.4: Radial Weir Gate

1.6.2. BYPASS MODIFICATION

The capacity of the existing Sacramento Bypass does not meet new regulatory standards for levee freeboard. The bypass was modeled for 900 and 1,800 foot expansions to analyze the increase in capacity and levee freeboard. A curved bypass alignment at the Yolo Bypass connection for the 900 foot was also considered.

1.6.2.1. 900 FOOT EXPANSION

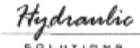

The 900 foot design alternative evaluates the expansion of the Sacramento Bypass to the north, including levee relocation, and extension of the Sacramento Weir structure by 900 feet. The extension of the weir structure includes the addition of new weir bays using the same dimensions of the existing weir.

1.6.2.2. 1,800 FOOT EXPANSION

The 1,800 foot design alternative evaluates the expansion of the Sacramento Bypass to the north, including levee relocation, and extension of the Sacramento Weir structure by 1,800 feet. The extension of the weir structure includes the addition of new weir bays using the same dimensions of the existing weir.

1.6.2.3. CURVED BYPASS CONNECTION

The expansion of the bypass will encroach onto the existing landfill located along the north levee near the confluence with the Yolo Bypass. Encroachment into the landfill requires the need to move or cap the landfill. Each of the two options will be costly to implement. A third alternative of avoidance was considered by redirecting the Sacramento Bypass with curvature away from the landfill. This alternative is only being reviewed for the 900 foot expansion.

2.0. HYDROLOGY

2.1 DESCRIPTION OF WATERSHED

The Sacramento River Basin, which can be seen in Figure 2.1, lies between the Sierra Nevada and Cascade Ranges in the east and the Coast Range and Klamath Mountains in the west. Its source waters rise in the volcanic plateaus and ranges of northern California as three rivers: the Upper Sacramento, McCloud, and Pit. These three rivers join in the waters of Lake Shasta, a 4.5 million acre-foot reservoir formed by Shasta Dam. From the dam the Sacramento River winds approximately 30 miles south through the foothills between Redding and Red Bluff. Many small and moderate-sized tributaries join the river from both east and west, including Clear, Cottonwood, Cow, and Battle Creeks. At Red Bluff a large portion of its flow is diverted into canals that deliver irrigation water to agriculture south in the Sacramento Valley. The Sacramento River continues to meander south, where it is joined by Antelope, Mill, and Deer Creeks in eastern Tehama County, and by Stony and Big Chico Creeks south of Chico. Butte Creek merges with the Sacramento River near Colusa and the Sutter Buttes, a group of isolated volcanic hills in the middle of the Sacramento Valley. The Sacramento River is joined by its largest tributary, the Feather River, at Verona. About 10 miles downstream, the Sacramento River flows through the city of Sacramento and receives the American River, its second largest tributary. At the confluence with the American River, the Sacramento River divides to provide a main stem and a navigable segment for cargo ships, which is identified as the Sacramento Deep Water Ship Channel. Both waterways rejoin in the estuary of the Delta near Rio Vista. The mouth of the Sacramento River is at Suisun Bay near Antioch, where it combines with the San Joaquin River. The Sacramento River, now nearly a mile wide at its mouth, flows into San Francisco Bay and finally joins the Pacific Ocean under the Golden Gate Bridge in San Francisco. (Lundgren, 2013)

The Sacramento River Basin provides drinking water for most of Northern and Southern California residents, as well as hundreds of wildlife species and agricultural industries. Over two

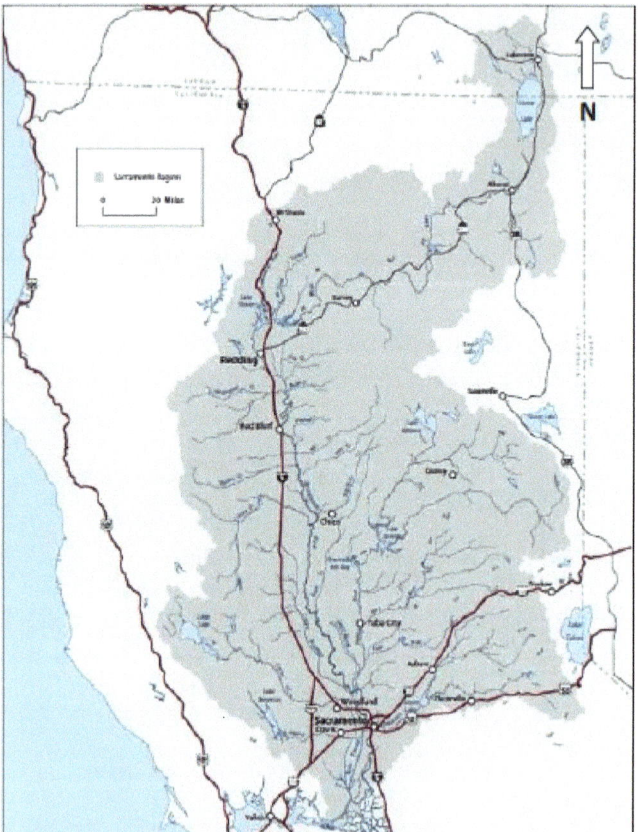

Figure 2.1: Sacramento River Basin

million people live in the Sacramento River Watershed and its urban land use is concentrated around the cities of Sacramento, Redding, Chico, and Red Bluff.

In the high mountain areas of the Sierra Nevada the average precipitation is about 85 inches per year, in the Redding area the average precipitation is 40.9 inches, and in the Sacramento area the average precipitation is 17.2 inches (US Army Corps). The average annual precipitation for the entire Sacramento River Basin is 35.9 inches, most of which falls as rain or snow during November through March.

2.2. DESCRIPTION OF SACRAMENTO RIVER AND PROJECT LOCATION

The Sacramento Weir and Bypass is located along the right bank of the Sacramento River approximately three miles upstream of the City of West Sacramento. The mouth of the American River is located about two miles downstream at Discovery Park. The Sacramento River is bounded by agricultural land; however, urban areas surround the project site: the Natomas region to the north and the cities of West Sacramento and Sacramento to the southeast. Interstate 80 crosses the Sacramento River to the southeast of the project site near the City of West Sacramento. Interstate 5 follows the east bank of the Sacramento River through the City of Sacramento. The Garden Highway travels along the east bank of the Sacramento River at the project location. Old River Road travels along the west bank, utilizing the existing bridge structure at the project site. The weir and bypass are located approximately four miles upstream of the Tower Bridge. Refer to Figure 2.2 for a map of the project site.

2.2.1 CHANNEL AND STRUCTURE GEOMETRY

The Sacramento River is a natural river channel that flows north to south through urban and agricultural lands. The width of the Sacramento River at the project site is approximately 900 feet. The Sacramento River is directed through the use of levees to prevent diversion and flooding into the neighboring properties. The levees were constructed to protect local interests but were later incorporated into the SRFCP and portions were improved by the USACE. The Sacramento River has a trapezoidal shape with an approximate depth of 25 feet at the Sacramento Bypass.

The Sacramento Bypass is a 1.7 mile channel bounded by the California Highway Patrol Academy to the south, privately owned agricultural land to the north, the Sacramento River to the east, and the Yolo Bypass to the west. The Sacramento Bypass is an 1,800 foot wide trapezoidal channel used to divert flow from the Sacramento River to the Yolo Bypass. The channel is directed with the use of levee structures to protect the properties to the north and south. The levees are constructed with earthen material and protected from erosion and water seepage with a concrete surface. The base of the Sacramento Bypass is vegetated to varying degrees with trees, bushes, and grasses.

The Sacramento Weir is a 1,920 foot long concrete structure consisting of 48 weir bays mounted with wooden needle weir gates. Each bay is 38.25 feet wide and has a weir elevation of 24.75 feet when the gates are opened and 31.0 feet when the gates are closed. Each gate consists of 38 wooden needles with a dimension of 4 inches thick by 1 foot wide by 6 feet long. The needles are hinged at the bottom to collapse in the direction of the flow acting as a broad crested weir.

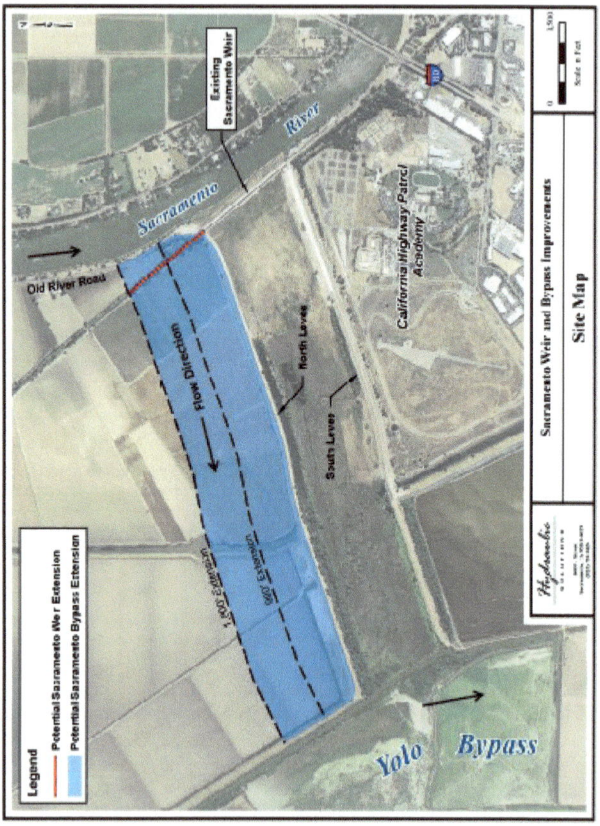

Figure 2.2: Site Map

2.2.2 FLOOD CONTROL

The Sacramento River Flood Control Project was designed with the understanding that runoff from many of the storm events experienced in the Sacramento River watershed cannot be contained within the banks of the river or be contained within a levee system without periodically flooding adjacent property. Thus, the SRFCP was designed to occasionally spill through a system of weirs and flood relief structures into adjacent basins. These basins are designed to contain floodwaters and channel them downstream, to eventually be conveyed back into the Sacramento River near Knights Landing and Rio Vista. There are ten overflow structures in the SRFCP: six weirs, three flood relief structures, and an emergency overflow roadway. Of these structures the Tisdale Weir releases overflow waters of the Sacramento River into the Sutter Bypass via the Tisdale Bypass, and the Fremont Weir releases overflow waters of the Sacramento River, Sutter Bypass, and the Feather River into the Yolo Bypass. The Sacramento Weir diverts Sacramento River and American River floodwaters to the west, down the mile long Sacramento Bypass and into the Yolo Bypass. The Sacramento Weir's primary purpose is to protect the City of Sacramento from excessive flood stages in the Sacramento River channel downstream of the American River. The Sacramento Weir helps to control flood stages in the Sacramento River through the Sacramento and West Sacramento area. The design capacity of the Sacramento Weir is 112,000 cubic feet per second (cfs). (Sacramento Flood Control Project. 2010)

3.0. HYDRAULIC ANALYSIS

3.1. HYDRAULIC MODELING

For the purpose of this project, an existing one-dimensional hydraulic model was used to establish baseline conditions. The same model was modified to develop design alternatives. The Sacramento Basin model was provided by DWR and initially developed using HEC-RAS by the USACE as part of Sacramento and San Joaquin River Basins Comprehensive Study (Comp Study). The current comprehensive Sacramento Basin HEC-RAS model incorporates individually developed models, such as: Comp Study UNET Model, American River, Feather/Yuba River, Sacramento River, Cache Creek and Tisdale Bypass. Initially the model was assembled having two parts: Upper and Lower Sacramento Basin. These two models were combined into one base model. Recalibration was performed every time a new reach was added to the model.

3.1.1. MODELING ASSUMPTIONS

Two Sacramento Bypass alternative hydraulic model scenarios that reflect the 900 and 1,800 foot expansion of the bypass were performed with the following assumptions:

- Flow in the channel is subcritical open channel flow;
- 100-year and 200-year unsteady flow hydrographs developed by USACE were used during simulation runs;
- Storm centering technique was used by USACE for determination of locations for inflow hydrographs;
- Flood flows in the Yolo Bypass may contribute to fluctuations in water surface elevation (WSEL) especially downstream of the Sacramento Bypass;

- During the simulation, flows in the channel interact with boundary condition structures, such as levees on the sides of the channel (once levees are overtopped, flows may go to adjacent storage areas), the Sacramento Weir, and the Yolo Bypass;
- Once opened, gates at the Sacramento Weir are modeled to remain fully open during the simulation runs;
- Once the weir is overtopped, flows through the bypass follow a wide but direct path to the Yolo Bypass;
- Sacramento Basin model was calibrated by USACE in 2012, it is assumed that current roughness conditions are adequately reflected in the model;
- Manning's Roughness Coefficient "n" values ("n" values) for widened bypass areas were approximated using Google Earth imagery (October 2011) and assigned to all extended segments of cross sections in order to account for resistance to the flow in the bypass, however test runs with newly established "n" values showed no significant effect on WSEL of the Bypass. Initial "n" values are used for the extended portions of the cross sections;
- Expansion of the bypass will have little or no effect on flows into adjacent storage areas, thus geometry and stage-volume curves for those storage areas are not going to be recalculated.

3.1.2. MODELED CROSS SECTIONS

The Sacramento Bypass is defined in the model by 10 cross sections. Modeled cross sections are generally perpendicular to the direction of assumed flow. The cross sections are evenly spaced along the length of the channel having an average distance of 1,000 feet between them. Two additional cross sections were created in the model identical to the most downstream and upstream sections of the bypass. These copies are set 1 foot apart from original most upstream and downstream cross sections. This was done in order to use the original cross sections as boundaries for the channel. All cross sections extend beyond the landside toe of the levees on the north and south sides of the bypass. The extensions of cross sections vary from 300 to 500 feet on both sides. Those extensions are modeled as obstructed areas to prevent unnecessary flow conveyance calculations during the simulation runs. See Appendix C for a map of the model's cross section locations.

3.1.3. FIELD SURVEYS AND DATUM

Because extensive site visits were not permitted, traditional field surveys were not available. An aerial survey from March-April 2008 using Light Detection and Ranging (LiDAR) technology was used. The scope of the survey included the general vicinity of the project site including both levees of the Sacramento Bypass, both levees of the Sacramento River near the project site, portions of the Yolo Bypass, and approximately 3,000 feet of farmland to the north of the Sacramento Bypass. The LiDAR survey has a vertical accuracy of 0.6 feet. Accuracy is lower in wet areas, terrain with heavy brush and trees, and under structures. The average spacing of the LiDAR point data was 3.28 feet. The LiDAR data was used to create a 3D surface with Geographic Information System (GIS) software. Ten 3D polylines were constructed approximately 1,000 feet apart through the bypass to be used as cross sections in the model. Those cross sections were imported directly into HEC-RAS.

The survey's horizontal datum was the North American Datum of 1983 (NAD 83), California State Plane Zone II, U.S. Feet. The vertical Datum was North American Vertical Datum of 1988 (NAVD 88).

3.1.4. HYDRAULIC STRUCTURES

The Sacramento Weir is modeled as a lateral structure along the Sacramento River. This structure has a tail water connection to the upstream cross section of the Sacramento Bypass. The flow over the weir is the upstream boundary condition for the Sacramento Bypass. Boundary conditions are discussed in the next section.

3.1.5. MODEL BOUNDARY CONDITIONS

A description of the watershed and hydrology for the model are discussed in section 2 of this report. As indicated, models of the Sacramento Bypass and Sacramento Weir were evaluated in the context of the comprehensive model of the Sacramento River Basin. The upstream most boundary of this model is the Feather River; stage hydrograph at the confluence of Sacramento River with Suisun Bay serves as the downstream boundary for the system. The model includes hundreds of hydraulically significant structures and inflow locations. The inflow hydrographs were developed by USACE based on results of Hydrologic Engineering Center – Hydrologic Modeling System (HEC-HMS) models that were developed for the Sacramento River system. A more detailed discussion regarding development of inflow hydrographs and the calibration process is provided in the section 3.2 of this report. Although the proposed expansions of the Sacramento Bypass affect the entire river system, it is beyond the scope of this project to provide detailed boundary condition analysis of the whole river system.

For the purpose of this analysis, the boundary conditions of the Sacramento Bypass were evaluated in the original model, and later adjusted for the purpose of developing design alternatives. The upstream boundary of the channel is the Sacramento Weir, the hydraulic structure currently operated by DWR. Once opened, the gates are modeled to remain open throughout the duration of the event, allowing maximum flow conveyance through the bypass. The weir flow hydrographs for 100-year and 200-year events are presented in Figures 3.1 and 3.2. Stages at the Yolo Bypass serve as the downstream boundary for the channel. The downstream stages were determined by simulation runs of 100-year and 200-year events and presented in Figure 3.3. At the location of the confluence, simulated peak flow through the Yolo Bypass at the 200-year event is 633,481 cfs; the peak flow through the Sacramento Bypass at the same event is 152,213 cfs. The Yolo Bypass is more than 5 times wider than the Sacramento Bypass. Thus, due to its greater capacity, the Yolo Bypass controls the stages at the junction of two bypasses. On south and north sides of the bypass the flow is contained by levees that are modeled as lateral structures. Refer to Appendix A for tabulated data.

3.1.6. CHANNEL ROUGHNESS

Manning's "n" values representing the channel's roughness were estimated because extensive site visits were not permitted. For the preliminary model runs, "n" values were already established in the model and were used; 0.04 for the right and left overbank and 0.035-0.038 for the channel.

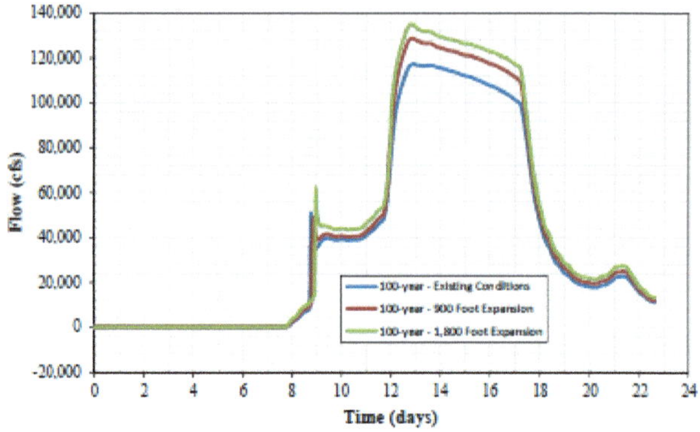

Figure 3.1: 100-year Weir Flow Hydrograph: Sacramento Bypass

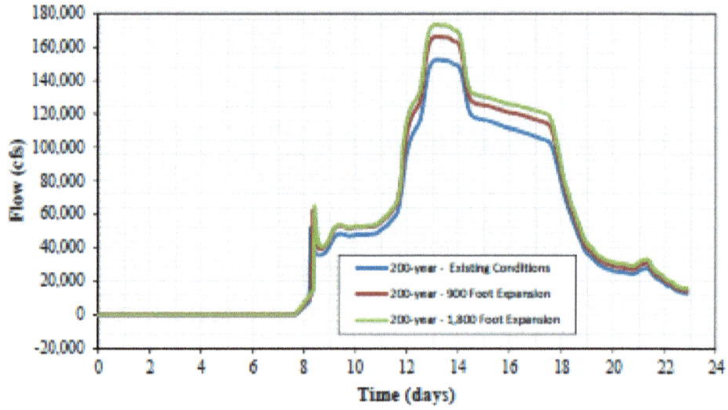

Figure 3.2: 200-year Weir Flow Hydrograph: Sacramento Bypass

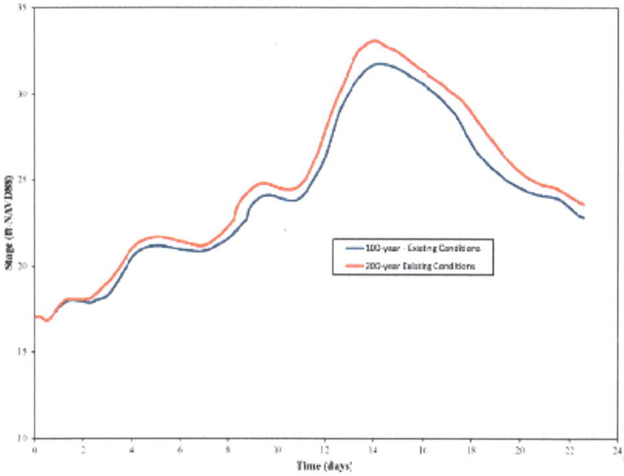

Figure 3.3: Yolo Bypass Stage Hydrograph at Sacramento Bypass Confluence: Existing Conditions

Hydraulic Solutions attempted to use a more detailed approach regarding accuracy of "n" values for additional model runs. The ArcGIS polylines used to create the 3D cross sections for the models were displayed over aerial imagery from October 2011. It was assumed maintenance of the Sacramento Bypass is typical from year to year and vegetation in the 2011 imagery was an accurate representation of current conditions. Areas of similar vegetation were measured along each cross section and appropriate "n" values were assigned based on table of values in the HEC-RAS reference manual (Appendix C). Initial test runs with the "n" values estimated from the imagery showed minimal effects on water surface elevations; therefore, hydraulic analysis for this report was performed with the existing model "n" values.

3.1.7. CHANNEL FLOW REGIME

Subcritical open channel flow is assumed throughout the bypass. Test runs of 100-year and 200-year flows with existing conditions confirmed that at the beginning of an event the Sacramento Bypass serves as a backwater storage area for Yolo Bypass. Once the gates of the Sacramento weir are open, the flow reverses its direction and the Sacramento Bypass becomes a tributary of the Yolo Bypass. Although reversal of the flow is expected, neither sudden change in the energy grade line nor change of the flow regime is anticipated in the channel, except the location immediately downstream of the weir, where the energy dissipater is located and a mixed flow regime is possible. As the channel is filled with water before major flow over the Sacramento weir is released, it is assumed that the main portion of the flow will travel along the whole width of the bypass being constricted only by levees on south and north sides until the confluence with the Yolo Bypass. There are no inline structures or other significant obstructions in the Sacramento Bypass that may constrict the flow.

3.2. MODEL CALIBRATION

The Sacramento River Basin Comp Study model provided by DWR is assumed to be calibrated. Verification of the accuracy of the calibration of the model is beyond the scope of the current project. The calibration of the model was performed by initial developers using gage stage data collected by USACE during the 1997 flood event. In the year 2000, the USACE had undertaken a task to develop hydrologic models for all major tributaries to the Sacramento River, as well as all major reservoirs. The USACE Hydrologic Engineering Center utilized HEC-HMS to perform the study. Geospatial Hydrologic Modeling System (HEC-GeoHMS) was used for sub-basin delineation, calculation of the physical characteristics of topography and estimation of hydrologic parameters. As a result, 33 individual HEC-HMS models for the main streams and the valley floor of the Sacramento and San Joaquin Rivers Basins were developed. The inflow hydrographs for the Sacramento River Basin HEC-RAS model were updated and recalibrated using gage stage data collected by USACE during the 2006 flood event. As indicated, in 2012 the USACE had undertaken the effort of recalibration of the model to reflect current roughness conditions of the streams and the valley floor.

3.2.1. MODEL LIMITATIONS

During the simulation runs of the model the following limitations should be taken into account:

- Calibration of the model was performed using gage stage data collected by USACE during the 1997 and 2006 flood events;
- The flow over the Sacramento Weir is modeled to be perpendicular to the Sacramento River;
- The energy dissipater located immediately downstream of the weir is not represented in the model;
- The Sacramento Weir is modeled as an open air weir with the assumption WSEL do not reach the lower chord of the bridge structure, pressurized flow is not anticipated.

3.3. MODELING SCENARIOS

The USACE and DWR recommended three different modeling scenarios for the Sacramento Weir and the Sacramento Bypass, which are the existing conditions, a 900 foot expansion of the weir and bypass, and 1,800 foot expansion of the weir and bypass. Analysis of the WSEL and levee freeboard for 100-year and 200-year flood events is required.

3.3.1. EXISTING CONDITIONS

The first modeling scenario is the existing condition model of the Sacramento Weir and Sacramento Bypass. This model represents the existing conditions of the Sacramento Weir and

Sacramento Bypass and its data was provided by the USACE. Plans were created for 100-year and 200-year flood events using the existing geometry file to run simulations for the existing conditions.

3.3.2. MODELING APPROACH FOR DESIGN ALTERNATIVES

The geometry of the existing condition model was modified to incorporate the 900 and 1,800 foot expansion of the bypass and weir. The dimensions for typical weir gate sections were kept the same as in the base model with a gate width of 38.25 feet and gate height of 18.73 feet for both design alternatives. Elevations at the weir invert were kept at 23.5 feet for all gates. Geometric characteristics of the existing and proposed weir are assumed to be the same. The proposed weir is modeled as a broad crested open air weir. As concrete is assumed to be the material for the construction of the new weir, the same weir coefficient (value of 3.2) for the existing weir was used for the analysis of both design alternatives.

The cross sections of the Sacramento Bypass were extended by 900 feet and 1,800 feet to create new profiles. The elevations of the North Levee of the bypass were removed from existing HEC-RAS cross sections leaving interpolated channel bed elevations based on the existing levee toes. The portions of the cross sections that correspond to the new proposed levee location were manually raised to simulate elevations of the existing North Sacramento Bypass Levee. The Sacramento River Levee (identified as lateral structure Sacramento River NCC to NEMDC 69.75 in the model) and Yolo Bypass Levee (identified as lateral structure Yolo Bypass KRLC to SacBP 47.48 in the model) were trimmed to accommodate the Sacramento Bypass expansion for both alternatives.

3.3.3. 900 FOOT SACRAMENTO BYPASS EXPANSION

The second modeling scenario is the model portraying the expansion of the Sacramento Weir and the Sacramento Bypass by 900 feet in the north direction. The existing condition model's geometry files were modified to incorporate the 900 foot expansion of the weir and bypass. The total length of the weir in the model is 2,810 foot for the 900 foot bypass expansion. The resulting total number of gates increased from 48 to 72 gates. The design capacity for the new weir was calculated using 2,333.3 cfs per gate. This value was derived from the design capacity of the existing weir, 112,000 cfs for 48 gates. Design capacity of the weir with 72 gates is 168,000 cfs. See Appendix B for sample calculations. The weir begins 20 feet. downstream of River mile 63.9979 for the 900 foot bypass expansion alternative.

3.3.4. 1,800 FOOT SACRAMENTO BYPASS EXPANSION

The third modeling scenario is the model portraying the expansion of the Sacramento Weir and the Sacramento Bypass by 1,800 feet in the north direction. The existing condition model's geometry files were modified to incorporate the 1,800 foot expansion of the weir and bypass. For this alternative, the length of the weir in the model is 3,689.6 feet. The total number of gates increased from 48 to 95 gates. The width of the 95th gate was reduced to 21.6 feet to fit within the proposed structure. Design capacity of the weir with 95 gates is 220,651 cfs. See Appendix B for sample calculations. The weir is modeled as a single lateral structure along the Sacramento River. The weir begins 700 feet. downstream of the River mile 64.2532 for the 1,800 foot bypass expansion alternative.

3.4. MODEL RESULTS

Water surface profiles for the Sacramento River, Yolo Bypass, Sacramento Bypass, and American River are presented and discussed in the following two subsections. Existing base geometry in the Sacramento River Basin Comp Study HEC-RAS model was used to generate WSEL and levee freeboard profiles using 100-year and 200-year flows. Geometries edited to incorporate 900 and 1,800 foot expansion of the Sacramento Bypass were used during the simulation runs to generate water surface profiles with the same 100-year and 200-year flows. Water surface profiles under the existing condition and the proposed 900 and 1,800 foot expansion alternatives were created and are discussed later in this section.

The freeboard analysis for the Sacramento Bypass is presented in subsection 3.4.3. The scope of the project required analysis of the levee freeboard within the Sacramento Bypass. Additional levee freeboard analysis was performed for the Sacramento River corresponding to the Sacramento downtown area (Comp Study RMs 57 to 71). This effort was undertaken in order to analyze the effect of expanding the Sacramento Bypass on freeboard for other critical locations within the Sacramento River system.

3.4.1. 100-YEAR FLOOD EVALUATION

Sacramento Bypass

Under the existing conditions, modeled maximum WSEL for the Sacramento Bypass varies between 31.76 and 34.28 feet and can be seen in Figure 3.4. Neither levee is overtopped during a 100-year event.

The expansion of the bypass by 900 feet resulted in a drop of maximum WSEL by almost 1 foot at the upstream cross section. It resulted in a rise of WSEL by a margin of 0.1 feet (1.2 inches) at the downstream cross section of the bypass when compared to the existing conditions. The range of WSEL with this design alternative varies between 31.86 and 33.27 feet.

Even with an anticipated higher peak flow of 134,744 cfs (See Figure 3.1 and Figure 3.4), the 1,800 foot expansion of the bypass leads to a further drop of 1.47 feet to the maximum WSEL at the upstream cross section. However, as was the case with the 900 foot expansion design alternative, the increase of maximum WSEL at the downstream cross section by a total of 0.16 feet (1.9 inches) is observed when compared with the existing conditions. The range of WSEL with the 1,800 foot expansion design alternative varies between 31.92 and 32.81 feet.

It can be seen in Figure 3.4 that both design alternatives resulted in an overall reduction of maximum WSEL through the majority of the Sacramento Bypass. Beginning at approximately River Mile (RM) 0.27 of the bypass, the maximum WSEL at the downstream cross sections were higher as a result of expanding the bypass. River Mile location maps can be found in Appendix C. The rise does not exceed 2 inches, thus levee overtopping is still not a concern. Since the model results may look counterintuitive regarding the drop of WSEL at the upstream and increase of WSEL at the downstream of the bypass, Hydraulic Solutions provides a possible explanation of the model behavior and WSEL results in section 3.4.2.Sacramento Bypass subsection.

Sacramento River
The plot of the Sacramento River 50 mile segment stretching from downstream of the Freeport Bridge to Knights Landing is presented in the Figure 3.5. With existing Sacramento Bypass geometry, levees along Sacramento River are overtopped in a few locations of this segment of the river. At RM 86.25, the WSEL is higher than the top of the levee (TOL) by 0.1 feet. The overflow with existing conditions is occurring at the storage area modeled as SA-RD1500; this area is labeled in Figure 3.5 as mostly agricultural land. The total overflowed volume is 185.82 acre-feet.

At RM 76.5016, the WSEL is higher than the TOL by 0.76 feet. The controlled overflow is occurring at this location to the Natomas Basin. The drop of TOL at the RM 60.8813 is due to the confluence of Sacramento and American rivers. In Figure 3.5, the drop of TOL at RM 59.505 is indicated, which may indicate overtopping at this location. Closer examination of this lateral structure however reveals that elevations of the TOL were substituted with elevations of the floodwall. Under the description of the lateral structure in the model it is indicated that the correction was made on 1/11/2011. The model indicates that there is no flow over the lateral structure that corresponds to RM 59.505. The lowest elevation of the floodwall at this location is 37.34 feet. This elevation will be used for the purpose of overtopping analysis when considering design alternatives.

It can be seen in Figure 3.5 that both design alternatives of the expansion of the Sacramento Bypass have marginal effect on WSEL upstream of the I-5 freeway. Starting from RM 110.5 of Colusa-Feather reach, a progressive drop in WSEL along the Sacramento River will occur. At RM 86.25 the difference between WSEL and TOL is reduced by less than 1 inch considering both design alternatives and overflow is still occurring at this location. The total overflowed volume is reduced to 120.9 acre-feet with the 900 foot expansion and 93.23 acre-feet with 1,800 foot expansion. At RM 76.5016, WSEL is higher than the TOL by 0.76 feet. The controlled overflow is assumed at this location to flow to the Natomas Basin; however the model does not indicate an overflow volume. As indicated in the previous paragraph, the TOL at RM 59.505 is substituted with the minimum elevations of the floodwall. Considering a significant drop in WSEL at this location by 0.91 feet with the 900 foot expansion and by 1.42 feet with the 1,800 foot expansion, overtopping is not occurring. The model indicated no flow over the corresponding lateral structure.

Although the freeboard analysis for the Sacramento River is beyond the scope of the current project, Hydraulic Solutions believes it is important, and must be considered. As indicated before, both design alternatives regarding the Sacramento Bypass expansion have notable effect on WSEL downstream of the Sacramento Weir. An example of such analysis was performed for the segment of the Sacramento River between RMs 57 and 71. The reach corresponds to the downtown area of Sacramento and City of West Sacramento. Analysis for the east and west levees are shown in Figures 3.6 and 3.7, respectively. Analysis was performed considering 100-year event flows. Upon approval, Hydraulic Solutions is prepared to perform analysis for different segments of the river for a 100-year event, as desired by the client. Examination of the previously discussed water surface profiles concludes that proposed design alternatives would help to meet the freeboard requirement of 3 feet for both east and west levees.

Yolo Bypass

According to the results obtained from unsteady state 100-year flow simulations, the predicted WSEL shows no significant differences between the existing conditions and the 900 and 1,800 foot design alternatives. Figure 3.8 is a graphic representation of the model data.

The 900 foot Sacramento Bypass expansion alternative provides an upstream maximum WSEL of 11.21 feet. When compared to the existing maximum WSEL of 11.22 feet, there is a resulting WSEL change of 0.01 foot (0.12 inches). The downstream maximum WSEL is 40.34 feet compared to the existing maximum WSEL of 40.37 feet and resulting in a WSEL change of 0.03 foot (0.36 inches). The results show the largest differences between the maximum WSEL for the 900 foot design alternative and the existing maximum WSEL is only 1.09 inches. Supporting data for Figure 3.8 can be found in Appendix A. There are locations where overtopping occurs but these locations are bounded by agricultural land and do not appear to be life threatening. The 900 foot expansion of the weir has no significant effect on WSEL or overtopping reduction.

The 1,800 foot Sacramento Bypass expansion alternative provides an upstream maximum WSEL of 11.20 feet. When compared to the existing maximum WSEL of 11.22 feet, there is a resulting WSEL change of 0.02 feet (0.24 inches). The downstream maximum WSEL is 40.32 feet compared to the existing maximum WSEL of 40.37 feet and resulting in a WSEL change of 0.05 feet (0.6 inches). The results show the largest differences between the maximum WSEL for 1,800 foot design alternative and the existing maximum WSEL is only 1.92 inches. There are locations where overtopping occurs, but these locations are bounded by agricultural land and do not appear to be life threatening. The 1,800 foot expansion of the weir also has no significant effect on WSEL reduction.

As expected, the results show that both alternatives have no significant effect on WSEL at upstream or downstream locations of the Yolo Bypass. The minimal effect is due to the large capacity of the Yolo Bypass relative to the additional conveyance of the Sacramento Bypass. However, it should be noted that at the confluence of two bypasses the WSEL increases slightly due to increased flow of the Sacramento Bypass anticipated with 900 and 1,800 foot design alternatives.

American River

Figure 3.9 graphically represents the American River WSEL under a 100-year design flood for existing conditions and both 900 and 1,800 foot Sacramento Bypass expansion alternatives. The figure shows that flows with the existing conditions, 900 and 1,800 foot expansions will not overtop the existing north and south levees of the American River.

The WSEL with the 900 foot expansion is lower than WSEL observed with the existing conditions by 0.91 feet at the junction of the American and Sacramento Rivers (refer to Appendix Table A-9 for river station = 0.12 miles). The stage difference progressively decreases moving further upstream along the American River.

The WSEL with the 1,800 foot expansion is lower than WSEL observed with the existing conditions by 1.43 feet. at the junction of the American and Sacramento Rivers (refer to Appendix Table A-9 for river station = 0.12 miles). The stage difference progressively decreases moving upstream along the American River.

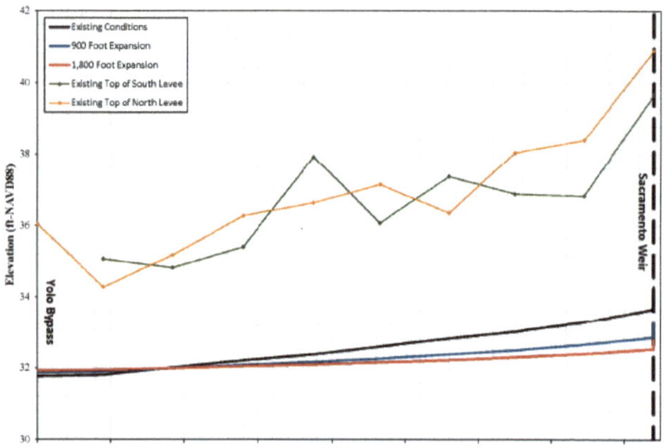

Sacramento Weir and Bypass Improvements: Hydraulic Analysis Report

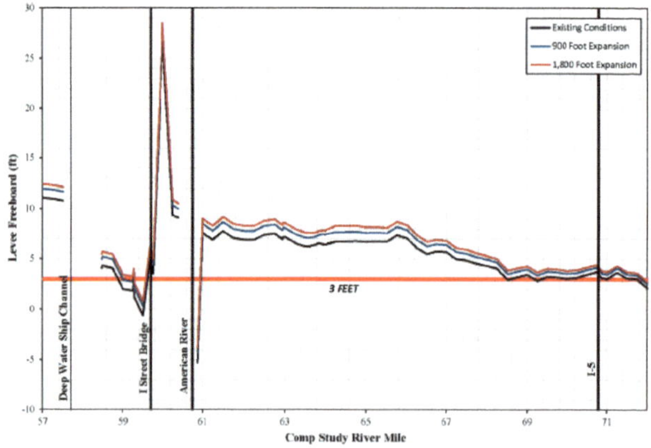

Figure 3.6: Sacramento River – East Levee Freeboard: 100-year
Downtown Area: I-5 to Deep Water Ship Channel

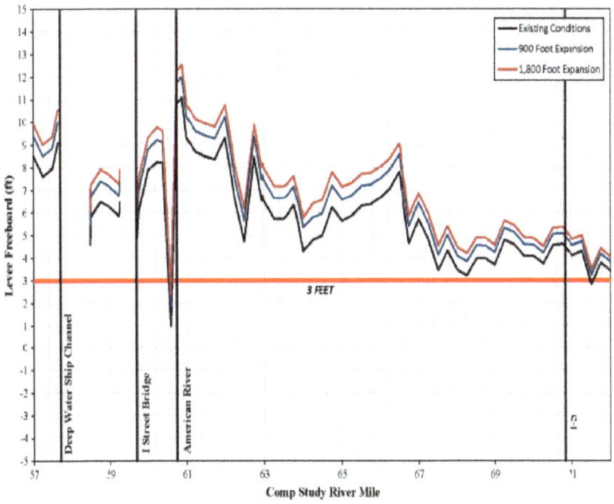

Figure 3.7: Sacramento River – West Levee Freeboard: 100-year
Downtown Area: I-5 to Deep Water Ship Channel

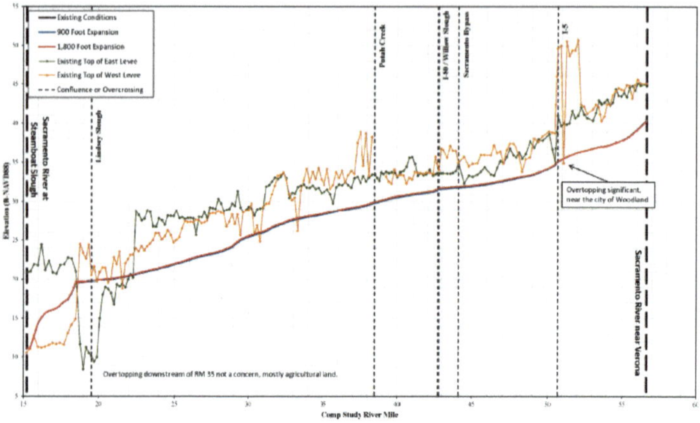

Figure 3.8: 100-year Water Surface Profile: Yolo Bypass

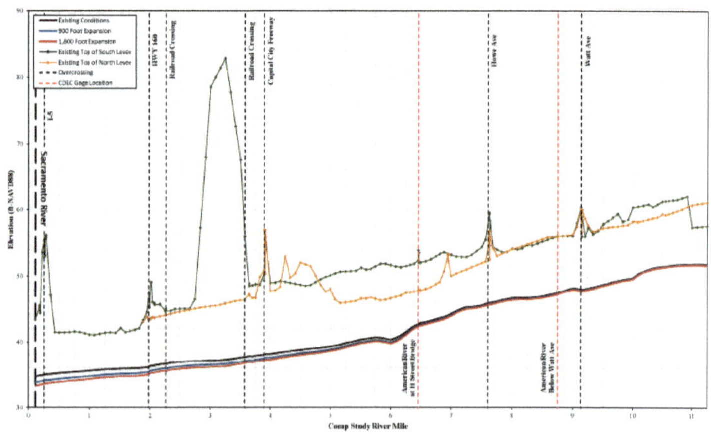

Figure 3.9: 100-year Water Surface Profile: American River

3.4.2. 200-YEAR EVALUATION

Sacramento Bypass
With existing conditions, the model produces WSELs that vary between 33.06 and 36.25 feet (Figure 3.10). Neither the North levee of the bypass nor the South levee is overtopped during a 200-year flood event.

The expansion of the bypass by 900 feet resulted in a drop of maximum WSEL by almost 1.23 feet at the upstream cross section, and a rise of WSEL by a margin of 0.13 feet (1.6 inches) at the downstream cross section of the bypass when compared to existing conditions. The range of WSEL with this design alternative varies between 33.19 and 35.02 feet (Figure 3.10).

Despite the fact that even higher flows through the bypass are anticipated during a 200-year event, the expectation of a drop in WSEL upstream of the bypass with an 1,800 foot expansion alternative was confirmed. A decrease in WSEL by a total of 1.82 feet at the upstream cross section was observed. However, as was the case with the 900 foot expansion alternative, the increase of maximum WSEL at the downstream cross section by the total of 0.21 feet (2.5 inches) was observed when compared to existing conditions. The range of WSEL with an 1,800 foot expansion alternative varies between 33.27 and 34.43 feet.

It can be seen in Figure 3.10 that both design alternatives result in an overall reduction of maximum WSEL throughout most of the Sacramento Bypass. As indicated in previous paragraphs, and as shown in the Figure 3.10, beginning at approximately RM 0.27 of the bypass, the maximum WSEL at the downstream cross sections were higher as a result of expanding the bypass. The rise is marginal (does not exceed 3 inches), thus levee overtopping is still not a concern.

Since the model results may look counterintuitive regarding the drop of WSEL at the upstream and the increase of WSEL at the downstream, Hydraulic Solutions performed the analysis of the modeling approach for the confluence of the Sacramento and the Yolo bypasses as well as the analysis of the interaction between channels during simulation runs. The results of that analysis are presented below.

Expansion of the Sacramento Bypass by 900 and 1,800 feet resulted in the change of boundary conditions for each design alternative. The peak flow over the Sacramento Weir during a 200-year event has increased from 152,536 cfs with existing conditions to 166,243 cfs with the 900 foot expansion, and to 173,310 cfs with 1,800 foot expansion. With an increase in flow of less than 15%, conveyance capacity of the bypass was increased by 50% and by 100% according to each design alternative. This is the major reason why the maximum WSEL in the bypass are lower when compared to the existing conditions. The expansion of the Sacramento Bypass by 900 or 1,800 feet contributes to the increase of inflow into the Yolo Bypass. This is a major reason why the increase of WSEL at the confluence is observed for both design alternatives.

Additional figures comparing 100-year and 200-year WSELs for the Sacramento Bypass for each design alternative can be found in Appendix C.

Sacramento River

The same reach of the Sacramento River used for the 100-year analysis was used for a 200-year simulation; the results can be seen graphically in Figure 3.11. With the existing Sacramento Bypass geometry, levees along the Sacramento River are overtopped in a number of locations. The locations of freeboard and overflow for the east and west levees of the Sacramento River are presented in Tables 3.1 and 3.2, respectively.

Table 3.1: Locations of Levee Overtopping – Sacramento River East Levee

River Mile	Maximum WSEL	Levee Freeboard	Overflow, $Q_{EAST,Base}$	Levee Freeboard 900 option	Overflow, $Q_{EAST,900}$	Levee Freeboard 1,800 option	Overflow, $Q_{EAST,1,800}$
	(ft)	(ft)	(cfs)	(ft)	(cfs)	(ft)	(cfs)
88.5	44.79	1.59		1.6		1.6	
86.75	44.65	-0.08	22245	-0.07	21261	-0.06	20672
86.25	44.7	-0.3		-0.28		-0.28	
79.0022	44.46	-0.25	79	-0.21	46	-0.19	31
78.7536	44.31	-0.25		-0.21		-0.18	

Table 3.2: Locations of Levee Overtopping – Sacramento River West Levee

River Mile	Maximum WSEL	Levee Freeboard	Overflow, $Q_{WEST,Base}$	Levee Freeboard 900 option	Overflow, $Q_{WEST,900}$	Levee Freeboard 1,800 option	Overflow, $Q_{WEST,1,800}$
	(ft)	(ft)	(cfs)	(ft)	(cfs)	(ft)	(cfs)
88.5	44.79	-0.33		-0.32		-0.32	
86.75	44.65	-0.37	52319	-0.36	50679	-0.35	49711
86.25	44.7	-1		-0.98		-0.98	
78.7536	44.31	-0.83		-0.79		-0.76	
78.2548	44.01	-0.12		-0.06		-0.03	
77.2512	43.4	-0.94		-0.85		-0.8	
77.0052	43.24	-0.55		-0.46		-0.4	
76.7497	43.1	-0.09		0.01		0.07	
76.5016	43.02	-1.8		-1.7		-1.64	
76.2542	42.88	-0.79	19549	-0.68	10409	-0.62	6801
75.7537	42.6	-1.22		-1.09		-1.01	
75.5022	42.46	-0.31		-0.16		-0.08	
75.2546	42.3	-0.73		-0.58		-0.48	
75.0012	42.21	-0.45		-0.29		-0.19	
74.5026	41.79	-0.58		-0.38		-0.26	
74.2531	41.74	-0.64		-0.43		-0.31	

As indicated during the analysis of 100-year flows, at RM 59.505 the elevation of the TOL was substituted with elevations of the floodwall. The model indicated there is no flow over the lateral structure that corresponds to this river mile.

Both design alternatives regarding the expansion of the Sacramento Bypass have little effect on WSEL upstream of the Sacramento Weir. Starting from RM 110.5 of the Colusa-Feather reach, a constant and progressive drop in WSEL along the Sacramento River is observed. The increase of freeboard and decrease of flow over corresponding lateral structures is demonstrated in the table. As indicated in the previous paragraph the TOL at the RM 59.505 is substituted with minimum elevations of the floodwall. Overtopping is not occurring at this location considering both design alternatives.

As was the case in the 100-year analysis, Hydraulic Solutions performed a 200-year freeboard analysis in the downtown Sacramento area and West Sacramento, between RMs 57 and 71. The freeboard analysis determined the design alternatives would help to meet the required freeboard of 3 feet for the downtown area; however the Natomas Region near I-5 does not have adequate flood protection. The freeboard deficiencies of the east and west levees of the Sacramento River near I-5 can be seen in Figures 3.12 and 3.13, respectively. As was stated earlier, further freeboard analysis for the Sacramento River can be performed at the request of the client.

Yolo Bypass

According to the results obtained from unsteady state 200-year flow simulations, the predicted WSEL shows no significant difference between the existing conditions, the 900 and 1,800 foot design alternative.

The maximum WSEL upstream of the bypass changes from 11.86 to 11.85 feet with no change in maximum WSEL downstream of the bypass with the 900 foot design alternative. Overtopping occurs at a few locations that can be seen on Figure 3.14. The largest difference in WSEL between the 1,800 foot design alternative and existing conditions is only 2.52 inches. The overtopping occurs at Lindsey Slough, around river station 18.474. The results show that both alternatives have no significant effect on the reduction of WSEL at the Yolo Bypass. This can be explained by the large capacity of the Yolo Bypass relative to the increased conveyance of the Sacramento Bypass.

American River

Figure 3.15 represents the WSEL at the American River during a 200-year event. Considering the existing conditions, 900 and 1,800 foot Sacramento Bypass design alternatives, overtopping of the existing north and south levees of American River is not anticipated. It should be noted however, at RM 11 on the existing south levee is near overtopping.

The WSEL with the 900 foot design alternative is lower than the WSEL with existing conditions by 1.04 feet at the confluence of the American and Sacramento Rivers (refer to Appendix Table A-9 for river station = 0.12 miles). WSEL with the 1,800 ft. expansion alternative is lower than WSEL with the existing conditions by 1.63 feet. at the same junction (refer to Appendix Table A-9 for river station = 0.12 miles). The difference in stage progressively decreases moving further upstream along the American River for both design alternatives.

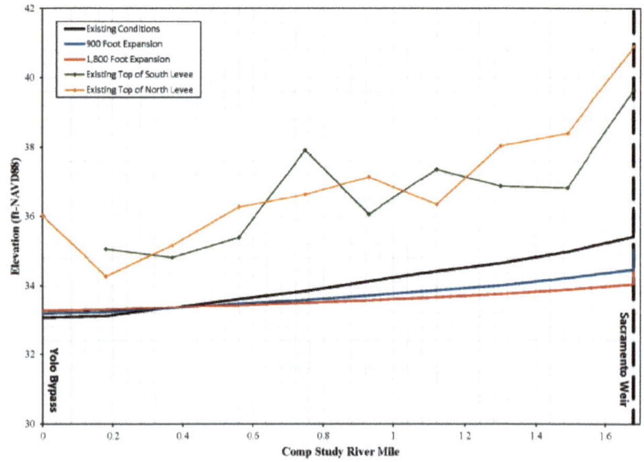

Figure 3.10: 200-year Water Surface Profile: Sacramento Bypass

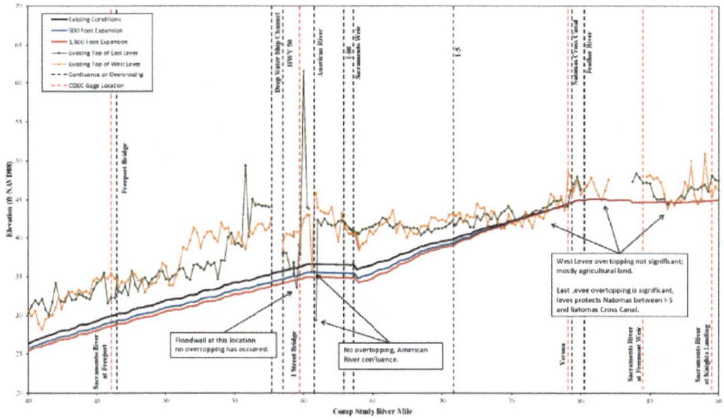

Figure 3.11: 200-year Water Surface Profile: Sacramento River

Sacramento Weir and Bypass Improvements: Hydraulic Analysis Report

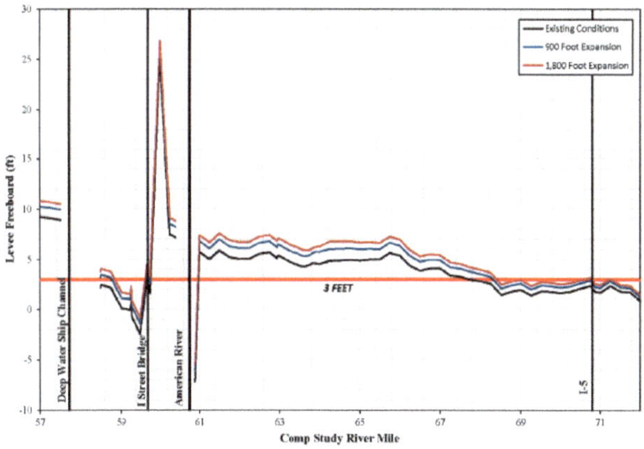

Figure 3.12: Sacramento River – East Levee Freeboard: 200-year Downtown Area: I-5 to Deep Water Ship Channel

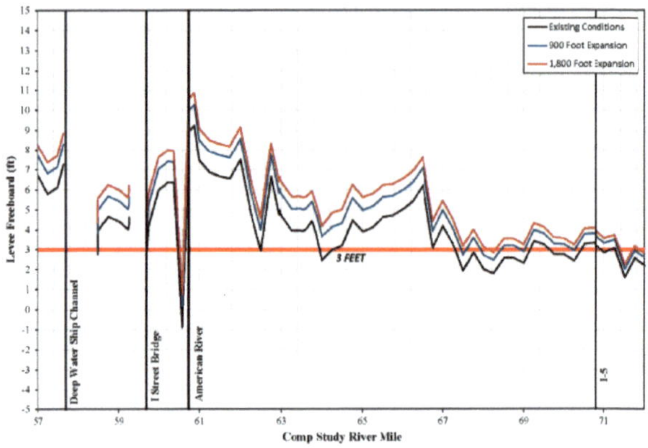

Figure 3.13: Sacramento River – West Levee Freeboard: 200-year
Downtown Area: I-5 to Deep Water Ship Channel

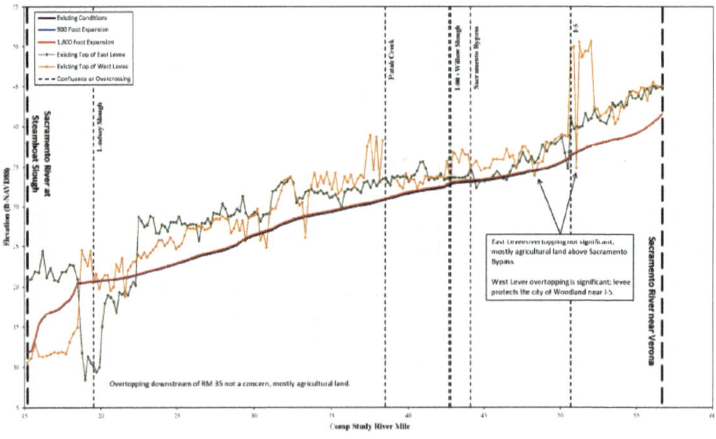

Figure 3.14: 200-year Water Surface Profile: Yolo Bypass

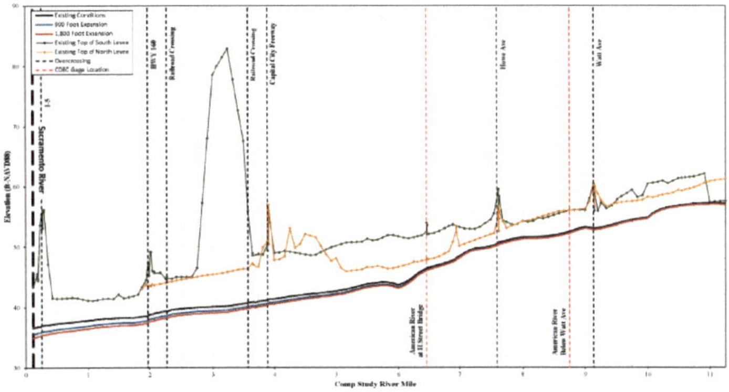

Figure 3.15: 200-year Water Surface Profile: American River

3.4.3 LEVEE FREEBOARD OF SACRAMENTO BYPASS

A freeboard analysis of both the north and south levees was performed on the Sacramento Bypass for existing conditions, 900 foot expansion, and 1,800 foot expansion. A 6 foot minimum freeboard was the required standard used for the analysis. Freeboard is the elevation difference between the TOL and the WSEL.

3.4.3.1. EXISTING CONDITIONS

Freeboard under existing conditions for both north and south levees of the Sacramento Bypass are graphically represented in Figures 3.16 and 3.17, respectively. During a 100-year event the north levee failed to meet the 6 foot freeboard requirement with the exception of a section of the bypass between RM 1.57 and the weir. The south levee failed to meet the freeboard requirement at all locations. During a 200-year event both north and south levees failed to meet freeboard requirements at all locations.

3.4.3.2. 900 FOOT EXPANSION

Predicted freeboard considering the 900 foot expansion for both north and south levees of the Sacramento Bypass are represented graphically in Figures 3.18 and 3.19, respectively. During a 100-year event the north levee failed to meet the 6 foot freeboard requirement with the exception of a section of the bypass between RM 1.51 and the weir. The south levee also fails to meet freeboard requirements with the exception of a section of the bypass between RM 1.62 and the weir. During a 200-year event both north and south levees failed to meet the freeboard requirement with the exception of a small section on the north levee between RM 1.65 and the weir.

3.4.3.3. 1,800 FOOT EXPANSION

Predicted freeboard considering the 1,800 foot expansion for both north and south levees of the Sacramento Bypass are represented graphically in Figures 3.20 and 3.21, respectively. During a 100-year event the north levee failed to meet the 6 foot freeboard requirement with the exception of a section of the bypass between RM 1.49 and the weir. The south levee also fails to meet freeboard requirements with the exception of a section of the bypass between RM 1.61 and the weir. During a 200-year event both north and south levees failed to meet the freeboard requirement with the exception of a small section on the north levee between RM 1.60 and the weir.

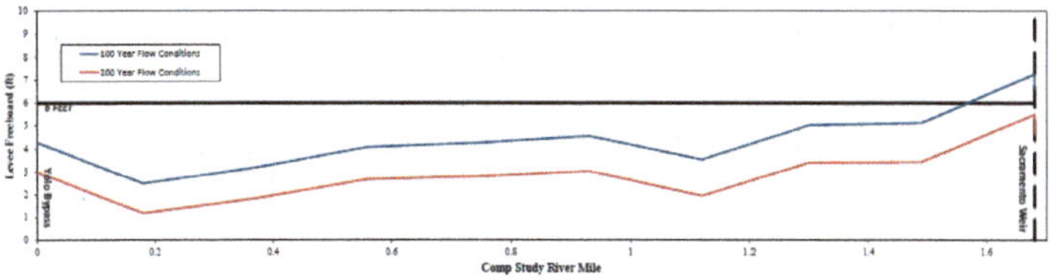

Figure 3.16: Sacramento Bypass North Levee Freeboard: Existing Conditions

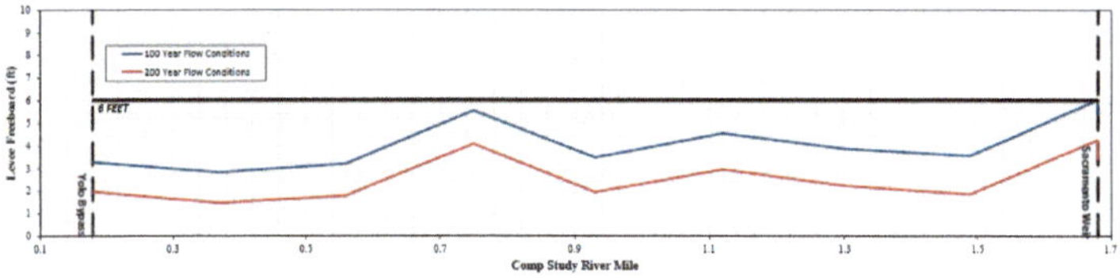

Figure 3.17: Sacramento Bypass South Levee Freeboard: Existing Conditions

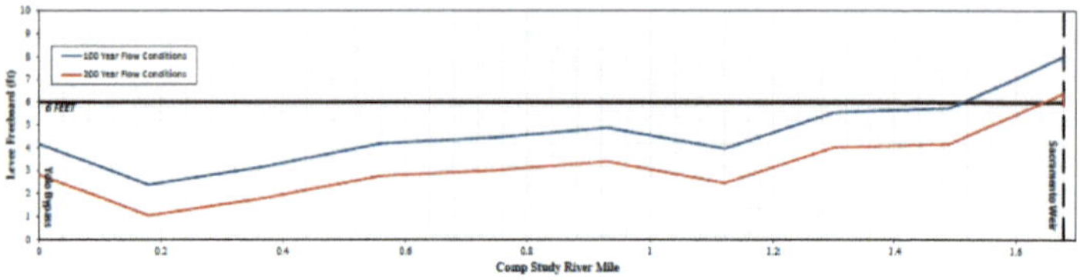

Figure 3.18: Sacramento Bypass North Levee Freeboard: 900 Foot Expansion

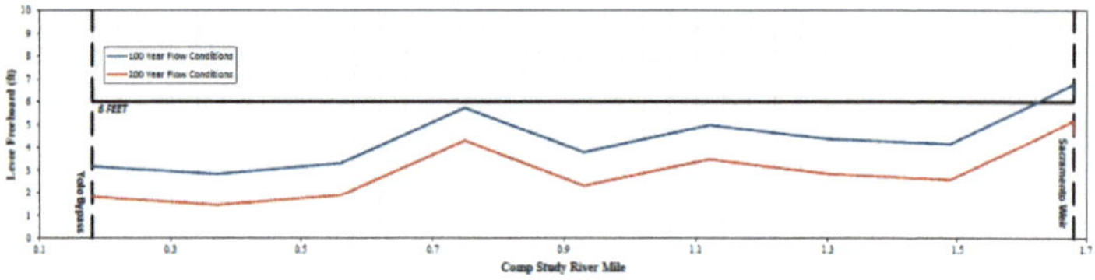

Figure 3.19: Sacramento Bypass South Levee Freeboard: 900 Foot Expansion

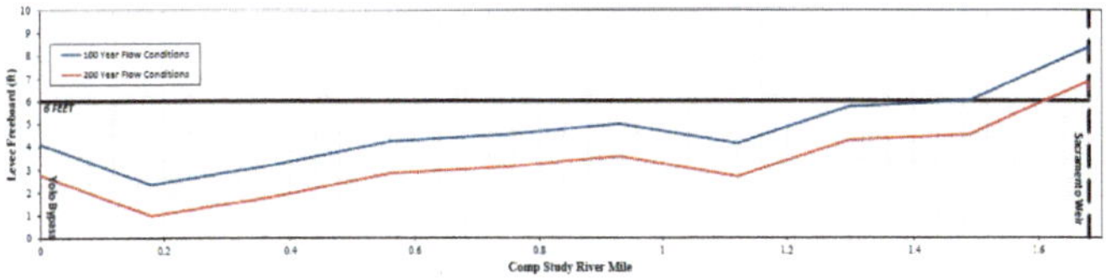

Figure 3.20: Sacramento Bypass North Levee Freeboard: 1,800 Foot Expansion

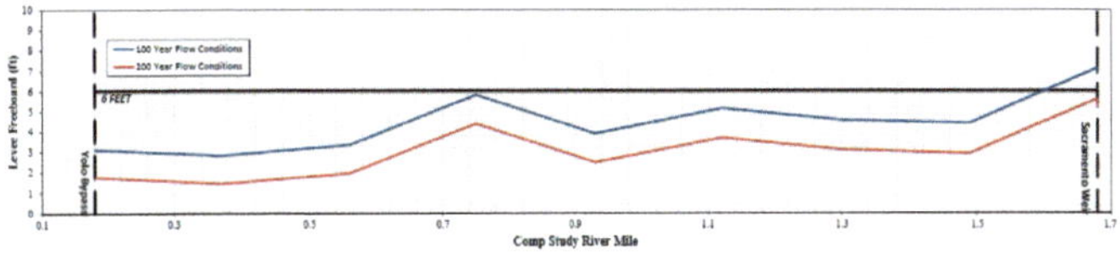

Figure 3.21: Sacramento Bypass South Levee Freeboard: 1,800 Foot Expansion

3.5. SCOUR ANALYSIS

Scour analysis of the weir and bridge structure will be performed by H_3Flo Consulting because the existing USACE model of the Sacramento Weir and Sacramento Bypass is not configured to perform a valid scour analysis. The existing USACE model has the weir configured as a lateral structure parallel to the river channel to enable the weir interaction between the Sacramento River and Sacramento Bypass. HEC-RAS requires a bridge model or inline structure to be configured perpendicular the channel to perform bridge scour calculations (reference HEC-RAS v4.1 Applications Guide: Chapter 11: Bridge Scour). A separate model would have to be developed to analyze scour specifically at the weir and bridge structure and it would be inappropriate to implement that scour model into the existing flood simulation model.

3.6. WEIR GATE ALTERNATIVES

The existing gate system uses wooden needle gates which are hinged at the bottom and held in place with a metal beam. The Department of Water Resources is interested automating the gates system to reduce the risks to health and safety of maintenance staff and increase efficiency of the weir system. Inflatable weir gate and radial weir gate systems were evaluated.

3.6.1. INFLATABLE WEIR GATE SYSTEMS

Automated inflatable weir gate systems were researched as an alternative to the manually operated weir gate currently in use. The inflatable weir gates consist of a series of steel gate panels supported on the downstream side by inflatable rubber air bladders. The bladders are attached to a concrete slab with stainless steel anchor bolts at six inches on center. The bladders are clamped over the anchor bolts and connected to the air supply pipes. The bladder hinge flaps are fastened to the gate panels. See Figure 1.3 for typical gate detail. The wedge-shaped profile of the inflatable weir gate system causes stable flow separation from the downstream edge of the gate without the vibration-inducing vortex shedding associated with simple rubber dams during overtopping. This results in a vibration-free operation and excellent controllability throughout a wide range of water elevations and gate positions.

The gates open and close through air pressure change within the bladders from the use of an air compressor. Total inflation occurs within approximately 30 minutes. The control system automatically maintains internal pressure and can be operated remotely from an office computer workstation with the addition of a modem and a phone line. A small building is required to house air compressors and system control equipment. Air piping will be required to cross the bypass and connect to the bladders at each gate.

3.6.2. RADIAL WEIR GATE SYSTEMS

Radial weir gates use a curved steel face plate mounted on radial arms that rotate on pivot joints anchored to the weir structure. The radial gate is considered a suitable alternative due to its simple design, relative light weight, and low hoist capacity requirements. The use of side seals eliminates the need for gate slots that are conducive to local low-pressure areas and possible cavitation.

Low lifting forces are required for radial gates, which typically uses wire rope attached at both ends, but can also operate with the use of hydraulic cylinders.

Radial gates are designed in the closed position with a minimum of 1 foot of freeboard above the normal upstream pool.

4.0 OPERATIONS AND MAINTENANCE REQUIREMENTS

4.1. BYPASS

Maintenance of the Sacramento Bypass, including levees, will remain the responsibility of the SMY. Maintenance responsibilities will continue to include, but are not limited to the following:

- Visual inspection of the bypass and levee structures for scour, sediment deposition, and vegetation.
- Removal of sediment and debris deposits within the stilling basin, as needed.
- Repair levees, as needed.
- Remove vegetation, as needed.

Typically maintenance will continue to occur outside the flood season to allow full inspection of the bypass when not in use and to limit health and safety hazards to SMY personnel.

4.2. WEIR GATES

Operations of the weir gates are user determined. It is expected that the timing of opening and closing the gates will continue to be based on the river stage of the Sacramento River at the I Street gage and weather forecasting as determined by the FOC. Since the gate will not require manual operation, DWR will need to determine if the FOC or SMY personnel will be responsible for the control of the automated weir gates. It is recommended that the FOC personnel maintain the controls to open and close the weir gates and the SMY perform all required maintenance of the weir gate.

The SMY will conduct general maintenance of the weir structure, including visual inspections of the structure and the removal of sediment deposits and debris from the stilling basin.

4.2.1. INFLATABLE WEIR GATE MAINTENANCE

Maintenance of the inflatable weir gates will require tightening the torque of the bladder clamps once a year to maintain bladder air pressure. Visual inspection of the gates should occur after major storm events to remove any debris that may cause damage to the bladder. With regular maintenance, the bladder life is 30 years.

4.2.2. RADIAL WEIR GATE MAINTENANCE

Maintenance of the radial weir gates will require visual inspection of all moving parts, including any rollers, hoists, pivot joints, etc. Maintenance of the radial weir gates may require the repair of replacement of some moving parts such as rollers, hoists, pivot joints, etc.

5.0 COST ANALYSIS

Estimated cost of the inflatable weir gate with a typical gate dimension of 38 feet. wide and 6 feet. tall was $114,000. The 1,800 foot expansion scenario provides a total of 95 gates. An assumed 25% increase to the overall gate cost for the materials and construction of the air piping, compressors, control equipment, and etcetera, provides a total gate cost of $13,537,500.

Estimated cost of the radial weir gate with a typical gate dimension of 38 feet wide and 6 feet tall was $60,000. The 1,800 foot expansion scenario provides a total of 95 gates. An assumed 25% increase to the overall gate cost for the materials and construction of the air piping, compressors, control equipment, and etcetera, provides a total gate cost of $7,125,000.

6.0 RECOMMENDATIONS AND CONCLUSIONS

6.1. SACRAMENTO BYPASS CONFIGURATION

Hydraulic Solutions recommends proceeding with the 1,800 foot expansion of the Sacramento Bypass. There are three major reasons that support this design alternative:

- Both 900 and 1,800 foot expansion alternatives lead to a significant increase in freeboard for urban levees, in the Sacramento downtown area in particular. However the increase in freeboard associated with the 1,800 foot expansion alternative is greater (in general by 0.5 feet) than the increase associated with the 900 foot expansion alternative. It can be seen in Figure 3.12 for east levee and in Figure 3.13 for the west levee of the Sacramento River that the 1,800 foot expansion helps meet the 3 foot freeboard requirement in the downtown area.

- Higher capacity of the Sacramento Bypass channel would provide conveyance of flows that are higher than those that were analyzed for 100-year and 200-year events. Hydraulic Solutions recommends to develop flow hydrographs for 500-year event and to perform hydraulic analysis with the flows anticipated during such event.

- As indicated in the Model Results section of this report, the Sacramento Bypass for both 100-year and 200-year evaluations had lower WSELs with the 1,800 foot expansion alternative, than with the 900 foot expansion alternative (Figure 3.4 and figure 3.10). Projected lower WSEL with the 1,800 foot expansion will lead to an increase in freeboard for the north and south levees of the bypass. Detailed recommendations regarding levee improvements are provided in section 5.3 of this report.

6.2. WEIR GATE RECOMMENDATION

Hydraulic Solutions recommends the use of inflatable weir gates for the Sacramento Weir extension. The inflatable weir gates have minimal structural requirements, requiring a concrete surface for the anchor bolts. The radial gates require more structural support to handle the additional weight of the gate added to the weir or bridge structure. The radial gates also have more moving parts requiring additional maintenance compared to the low maintenance of the inflatable weir gates.

6.3 RECOMMENDED LEVEE IMPROVEMENTS

The Sacramento Bypass levee freeboard results in section 3.4.3. prove that the Sacramento Bypass' north and south levees will not be able to meet the 6 foot freeboard requirement throughout most of the reach. Although the 900 and 1,800 foot design alternative increases freeboard initially at the Sacramento Weir, both still fail to meet the required freeboard for a 100-year and 200-year event. Therefore, Hydraulic Solutions recommends increasing the levee elevation based on the lowest freeboard in the reach. Table 6.1 displays the lowest elevation for all levee freeboard plots.

Table 6.1: Lowest Freeboard for South/North Levee with Existing/900-ft/1,800-ft

	South Levee Freeboard (ft)		North Levee Freeboard (ft)	
	100-year	200-year	100-year	200-year
Existing	2.80	1.42	2.50	1.15
900-ft Expansion	2.82	1.45	2.38	1.0
1,800-ft Expansion	2.82	1.42	2.30	1.0

Table 6.2 displays the minimum required levee elevation increase for all related conditions.

Table 6.2: Minimum Required Levee Elevation Increase

	South Levee Elevation Increase (ft)		North Levee Elevation Increase (ft)	
	100-year	200-year	100-year	200-year
Existing	3.20	4.58	3.50	4.85
900-ft Expansion	3.18	4.55	3.62	5.00
1,800-ft Expansion	3.18	4.58	3.70	5.00

With a 200-year flood design and based on the highest minimum elevation increase of 5.00 feet, Hydraulic Solutions recommends both the north and south levees' elevation to be increased by 5 feet.

7.0. REFERENCES

County of Sacramento, "Lower Sacramento River Information" and "Lower American River Information" Published January 15, 2013 Department of Water Resources Web. 17 March 2013: <http://www.sacflood.org/fludlinks.htm>.

Cowin, Mark W. *Urban Level of Flood Protection Criteria*. Rep. Department of Water Resources, State of California, Apr. 2012. Web. 17 Mar. 2013.

Domagalski, Joseph, Peter Dileanis, Jennifer Shelton, and Larry Brown. "Sacramento River Basin, National Water Quality Assessment Program." *Sacramento River Basin, National Water Quality Assessment Program*. U.S. Geological Survey, 14 June 2011. Web. 5 Mar. 2013. <http://ca.water.usgs.gov/sac_nawqa/study_description.html>.

Domagalski, Joseph L. "Water Use in California." *Water Quality in the Sacramento River Basin, California, 1994 - 98*. Denver: US Geological Survey, Information Services, 2000. 5. Print.

Engineering and Design - Planning and Design of Navigation Dams. Rep. no. EM 1110-2-2607. Sacramento: USACE, 1995. Print.

Kukas, Greg. *Post-Authorization Change Report and Interim General Reevaluation Report*. Rep. Sacramento: USACE, 2010. *American River Watershed*. Sacramento Area Flood Control Agency, July 2010. Web. 17 Mar. 2013.
<http://www.safca.org/documents/NLIP%20main%20page%20stuff/2010June.PACR%20American%20River%20Watershed%20Commpon%20Features%20Project%20Natomas%20Basin%20Sacramento%20&%20Sutter%20Counties,%20Ca/NPACRHydraulicAppCJuly2010s.pdf>

Lundgren, Chuck, Dennis Bowker, and Holly Jorgensen. "Life in the Watershed." *Sacramento River Watershed Program*. Battle Creek Watershed Conservancy, n.d. Web. 9 March 2013. <http://www.sacriver.org/aboutwatershed/roadmap/sacramento-river-basin>.

"PRATT | Henry Pratt Company." *Valve Manufacturer*. N.p., n.d. Web. 17 Mar. 2013. <http://www.hydrogate.com/>.

Russo, Mitch, comp. *Sacramento River Flood Control Project*. Tech. Sacramento: Department of Water Resources, 2010. Print.

Sacramento River Watershed Map. 2013. Photograph. Sacramento.

Types of Navigation Dam Structures. Rep. USACE, 31 July 1995. Web. 10 Mar. 2013. <http://140.194.76.129/publications/eng-manuals/EM_1110-2-2607/c-5.pdf>.

USACE. 1955. Supplement to Standard Operation and Maintenance Manual. Sacramento River Flood Control Project 1-7, U.S. Army Corps of Engineers, Sacramento, California Web. 10 March 2013: <https://sacct.csus.edu>

USACE. Technical Studies Documentation. Appendix Synthetic Hydrology. Retrieved on March 12, 2013

Warner, John C., Gary W. Brunner, Brent C. Wolfe, and Steven S. Piper. *Bridge Scour*. Rep. USACE, Jan. 2010. Web. 17 Mar. 2013. <http://www.hec.usace.army.mil/software/hec-ras/documents/HEC-RAS_4.1_Applications_Guide.pdf>.

Zenobia, Kent. 2012. Regional Flood Management Planning. The 2012 Central Valley Flood Protection Plan (CVFPP), 1-24, Department of Water Resources, Sacramento, California Web. 10 March 2013: <https://sacct.csus.edu>

8.0. APPENDICES

APPENDIX A. TABLES

Table A-1: Sacramento Bypass Water Surface Profile: Existing Conditions

River Station (mi)	WSEL (ft)		Elevation (ft)		1957 Design		Channel Invert	
	100-Year Event	200-Year Event	South Levee	North Levee	River Station (mi)	WSEL (ft)	River Station (mi)	Avg Bottom Depth (ft)
0	31.76	33.06		36.03	0	29.3	0	13.15
0.000183	31.76	33.06		36.03	0.5	29.6	0.000183	13.37
0.18	31.8	33.11	35.06	34.28	1	30	0.18	14.21
0.37	32.01	33.37	34.82	35.17	1.49	31.4	0.37	14.71
0.56	32.2	33.62	35.39	36.27	1.68	34.5	0.56	14.13
0.75	32.37	33.85	37.91	36.63			0.75	16.16
0.93	32.59	34.13	36.06	37.14			0.93	15.89
1.12	32.82	34.42	37.36	36.35			1.12	15.77
1.3	33.01	34.65	36.87	38.04			1.3	17.01
1.49	33.28	34.98	36.82	38.4			1.49	18.47
1.67981	33.65	35.42	39.66	40.9			1.67981	19.55
1.68	34.28	36.25	39.66	40.9			1.68	19.18

Table A-2: Sacramento Bypass Water Surface Profile: 900 Foot Expansion

River Station (mi)	WSEL (ft)		Elevation (ft)		1957 Design		Channel Invert	
	100-Year Event	200-Year Event	South Levee	North Levee	River Station (mi)	WSEL (ft)	River Station (mi)	Avg Bottom Depth (ft)
0	31.86	33.19		36.03	0	29.3	0	13.15
0.000183	31.86	33.19		36.03	0.5	29.6	0.000183	13.37
0.18	31.89	33.23	35.06	34.28	1	30	0.18	14.21
0.37	31.98	33.35	34.82	35.17	1.49	31.4	0.37	14.71
0.56	32.08	33.40	35.39	36.27	1.68	34.5	0.56	14.13
0.75	32.17	33.6	37.91	36.63			0.75	16.16
0.93	32.26	33.73	36.06	37.14			0.93	15.89
1.12	32.38	33.88	37.36	36.35			1.12	15.77
1.3	32.49	34.02	36.87	38.04			1.3	17.01
1.49	32.66	34.23	36.82	38.4			1.49	18.47
1.67981	32.87	34.48	39.66	40.9			1.67981	19.55
1.68	33.27	35.02	39.66	40.9			1.68	19.18

Table A-3: Sacramento Bypass Water Surface Profile: 1,800 Foot Expansion

River Station (mi)	WSEL (ft)		Elevation (ft)		1957 Design		Channel Invert	
	100-Year Event	200-Year Event	South Levee	North Levee	River Station (mi)	WSEL (ft)	River Station (mi)	Avg Bottom Depth (ft)
0	31.92	33.27		36.03	0	29.3	0	13.15
0.000183	31.92	33.27		36.03	0.5	29.6	0.000183	13.37
0.18	31.94	33.3	35.06	34.28	1	30	0.18	14.21
0.37	31.99	33.38	34.82	35.17	1.49	31.4	0.37	14.71
0.56	32.05	33.45	35.39	36.27	1.68	34.5	0.56	14.13
0.75	32.1	33.52	37.91	36.63			0.75	16.16
0.93	32.16	33.59	36.06	37.14			0.93	15.89
1.12	32.22	33.68	37.36	36.35			1.12	15.77
1.3	32.3	33.77	36.87	38.04			1.3	17.01
1.49	32.4	33.9	36.82	38.4			1.49	18.47
1.67981	32.54	34.06	39.66	40.9			1.67981	19.55
1.68	32.81	34.43	39.66	40.9			1.68	19.18

Table A-4: Sacramento River - East Levee Freeboard: 100-year Downtown Area: I-5 to Deep Water Ship Channel

River Station (mi)	Freeboard (ft)		
	Existing	900-ft	1,800-ft
57.0026	11.04	11.93	12.44
57.2522	10.92	11.81	12.32
57.4972	10.73	11.63	12.14
57.6416			
57.8526			
58.0013			
58.2496			
58.4871	4.05	4.96	5.46
58.4892	4.02	4.93	5.43
58.5008	4.28	5.19	5.69
58.5029	4.28	5.19	5.69
58.5042	4.28	5.19	5.69
58.5063	4.28	5.19	5.69
58.5179	4.27	5.17	5.68
58.5199	4.27	5.18	5.68
58.7508	4.01	4.92	5.43
59.0028	1.96	2.87	3.38
59.2503	1.77	2.69	3.2
59.2708	2.58	3.5	4.01
59.2729	2.52	3.43	3.94
59.2858	2	2.91	3.42
59.2878	1.09	2	2.51
59.5005	-0.64	0.27	0.78
59.6782	4.71	5.63	6.15
59.6817	4.71	5.63	6.15
59.6838	4.71	5.63	6.15
59.6894			
59.6915	4.61	5.53	6.04
59.6963	4.46	5.38	5.89
59.7519	3.46	4.38	4.9
59.9979	27.08	28.01	28.53
60.2518	9.33	10.31	10.86
60.3926	9.05	9.96	10.48
60.3928	9.06	9.97	10.49

River Station (mi)	Freeboard (ft)		
	Existing	900-ft	1,800-ft
60.5957			
60.5958			
60.7522			
60.8811			
60.8813	-5.31	-4.39	-3.87
60.9982	7.59	8.51	9.03
60.9983	7.59	8.51	9.03
61.2547	6.85	7.77	8.29
61.4986	7.73	8.65	9.17
61.7455	7.04	7.96	8.48
62.0032	6.85	7.77	8.29
62.2564	6.85	7.77	8.29
62.4965	7.35	8.27	8.79
62.7519	7.49	8.41	8.93
62.9407	6.92	7.84	8.36
62.9428	6.92	7.84	8.36
62.9519	7.05	7.97	8.49
62.954	7.08	8	8.52
62.9596	7.14	8.06	8.58
62.9617	7.15	8.07	8.59
62.9708	7.16	8.08	8.6
62.9729	7.16	8.08	8.6
63.0067	7.13	8.05	8.57
63.2518	6.54	7.46	7.98
63.2696	6.52	7.44	7.96
63.2874	6.49	7.41	7.93
63.3052	6.47	7.39	7.91
63.323	6.44	7.36	7.88
63.3408	6.42	7.34	7.86
63.3586	6.39	7.31	7.83
63.3765	6.37	7.29	7.81
63.3943	6.35	7.27	7.79
63.4121	6.32	7.24	7.76
63.4299	6.3	7.22	7.74

River Station (mi)	Freeboard (ft)		
	Existing	900-ft	1,800-ft
63.4477	6.27	7.19	7.71
63.4655	6.25	7.17	7.69
63.4833	6.22	7.14	7.66
63.5012	6.21	7.12	7.64
63.5191	6.2	7.11	7.62
63.5371	6.19	7.09	7.6
63.5551	6.19	7.08	7.59
63.5731	6.19	7.09	7.59
63.5911	6.2	7.07	7.57
63.6091	6.21	7.08	7.58
63.6271	6.2	7.08	7.57
63.645	6.23	7.09	7.58
63.663	6.25	7.1	7.58
63.681	6.27	7.11	7.58
63.699	6.28	7.12	7.58
63.717	6.29	7.14	7.6
63.735	6.32	7.14	7.59
63.753	6.33	7.16	7.6
63.7718	6.36	7.2	7.63
63.7906	6.42	7.23	7.66
63.8095	6.44	7.27	7.69
63.8283	6.49	7.32	7.73
63.8471	6.51	7.34	7.76
63.866	6.5	7.38	7.78
63.8848	6.48	7.41	7.8
63.9037	6.47	7.42	7.83
63.9225	6.45	7.43	7.82
63.9413	6.43	7.44	7.84
63.9602	6.41	7.43	7.83
63.979	6.4	7.43	7.83
63.9979	6.37	7.42	7.82
64.2532	6.67	7.67	8.29
64.5033	6.69	7.67	8.27
64.7562	6.74	7.71	8.3

River Station (mi)	Freeboard (ft)		
	Existing	900-ft	1,800-ft
65.0046	6.65	7.59	8.16
65.2513	6.72	7.63	8.18
65.5045	6.67	7.55	8.07
65.7539	7.32	8.18	8.69
66.0073	7.03	7.87	8.37
66.2529	6.04	6.87	7.36
66.5012	5.45	6.23	6.7
66.7511	5.69	6.44	6.89
67.0013	5.65	6.37	6.79
67.2532	4.95	5.66	6.08
67.5018	4.82	5.51	5.91
67.7513	4.51	5.18	5.57
68.0022	4.29	4.92	5.29
68.2473	4.02	4.63	4.99
68.5059	2.9	3.48	3.82
68.7477	3.17	3.74	4.07
69.0004	3.39	3.94	4.26
69.2531	2.81	3.34	3.66
69.5003	3.21	3.74	4.04
69.7512	3.14	3.65	3.95
69.9983	3.01	3.51	3.79
70.2464	3.12	3.6	3.88
70.4988	3.41	3.89	4.16
70.7462	3.68	4.14	4.41
70.8909	3.03	3.49	3.75
70.9239	3.06	3.51	3.77
71.0034	2.98	3.42	3.68
71.2551	3.64	4.07	4.31
71.5027	3.05	3.46	3.7
71.7536	2.96	3.36	3.59
71.9985	2.12	2.51	2.73
72.253	1.8	2.16	2.37
72.5009	3.07	3.42	3.62

Table A-5: Sacramento River - West Levee Freeboard: 100-year Downtown Area: I-5 to Deep Water Ship Channel

River Station (mi)	Freeboard (ft)		
	Existing	900-ft	1,800-ft
57.0026	8.55	9.44	9.95
57.2522	7.61	8.5	9.01
57.4972	7.94	8.84	9.35
57.6416	9.09	9.99	10.55
57.8526			
58.0013			
58.2496			
58.4871	5.69	6.6	7.1
58.4892	4.58	5.49	5.99
58.5008	5.79	6.7	7.2
58.5029	5.8	6.71	7.21
58.5042	5.8	6.71	7.21
58.5063	5.81	6.72	7.22
58.5179	5.84	6.74	7.25
58.5199	5.85	6.76	7.26
58.7508	6.48	7.39	7.9
59.0028	6.24	7.15	7.66
59.2503	5.84	6.76	7.27
59.2708	6.06	6.98	7.49
59.2729	6.46	7.37	7.88
59.2858			
59.2878			
59.5005	6.44	7.35	7.86
59.6782			
59.6817			
59.6838			
59.6894			
59.6915	4.02	4.94	5.45
59.6963	4.59	5.51	6.02
59.7519	6.05	6.97	7.49
59.9979	7.86	8.79	9.31
60.2518	8.24	9.22	9.77
60.3926	8.2	9.11	9.63
60.3928	8.18	9.09	9.61

River Station (mi)	Freeboard (ft)		
	Existing	900-ft	1,800-ft
60.5957	0.98	1.89	2.41
60.5958	0.99	1.9	2.42
60.7522	10.81	11.74	12.25
60.8811	11.08	12	12.52
60.8813	11.08	12	12.52
60.9982	9.31	10.23	10.75
60.9983	9.31	10.23	10.75
61.2547	8.68	9.6	10.12
61.4986	8.49	9.41	9.93
61.7455	8.36	9.28	9.8
62.0032	9.32	10.24	10.76
62.2564	6.61	7.53	8.05
62.4965	4.72	5.64	6.16
62.7519	8.45	9.37	9.89
62.9407	6.53	7.45	7.97
62.9428	6.57	7.49	8.01
62.9519	6.7	7.62	8.14
62.954	6.73	7.65	8.17
62.9596	6.75	7.67	8.19
62.9617	6.72	7.64	8.16
62.9708	6.62	7.54	8.06
62.9729	6.6	7.52	8.04
63.0067	6.47	7.39	7.91
63.2518	5.72	6.64	7.16
63.2696	5.72	6.64	7.16
63.2874	5.72	6.64	7.16
63.3052	5.72	6.64	7.16
63.323	5.72	6.64	7.16
63.3408	5.72	6.64	7.16
63.3586	5.72	6.64	7.16
63.3765	5.72	6.64	7.16
63.3943	5.72	6.64	7.16
63.4121	5.72	6.64	7.16
63.4299	5.72	6.64	7.16

River Station (mi)	Freeboard (ft)		
	Existing	900-ft	1,800-ft
63.4477	5.72	6.64	7.16
63.4655	5.72	6.64	7.16
63.4833	5.72	6.64	7.16
63.5012	5.73	6.64	7.16
63.5191	5.75	6.66	7.17
63.5371	5.78	6.68	7.19
63.5551	5.81	6.7	7.21
63.5731	5.84	6.74	7.24
63.5911	5.89	6.76	7.26
63.6091	5.93	6.8	7.3
63.6271	5.97	6.85	7.34
63.645	6.03	6.89	7.38
63.663	6.08	6.93	7.41
63.681	6.14	6.98	7.45
63.699	6.18	7.02	7.48
63.717	6.22	7.07	7.53
63.735	6.29	7.11	7.56
63.753	6.33	7.16	7.6
63.7718	6.2	7.04	7.47
63.7906	6.1	6.91	7.34
63.8095	5.96	6.79	7.21
63.8283	5.85	6.68	7.09
63.8471	5.71	6.54	6.96
63.866	5.54	6.42	6.82
63.8848	5.36	6.29	6.68
63.9037	5.19	6.14	6.55
63.9225	5.01	5.99	6.38
63.9413	4.83	5.84	6.24
63.9602	4.65	5.67	6.07
63.979	4.48	5.51	5.91
63.9979	4.29	5.34	5.74
64.2532	4.79	5.79	6.41
64.5033	4.97	5.95	6.55
64.7562	6.23	7.2	7.79

River Station (mi)	Freeboard (ft)		
	Existing	900-ft	1,800-ft
65.0046	5.63	6.57	7.14
65.2513	5.85	6.76	7.31
65.5045	6.3	7.18	7.7
65.7539	6.37	7.23	7.74
66.0073	6.66	7.5	8
66.2529	7.06	7.89	8.38
66.5012	7.8	8.58	9.05
66.7511	4.64	5.39	5.84
67.0013	5.71	6.43	6.85
67.2532	4.78	5.49	5.91
67.5018	3.45	4.14	4.54
67.7513	4.33	5	5.39
68.0022	3.46	4.09	4.46
68.2473	3.22	3.83	4.19
68.5059	3.97	4.55	4.89
68.7477	3.97	4.54	4.87
69.0004	3.69	4.24	4.56
69.2511	4.78	5.31	5.63
69.5003	4.6	5.13	5.43
69.7512	4.09	4.6	4.9
69.9983	4.07	4.57	4.85
70.2464	3.74	4.22	4.5
70.4988	4.57	5.05	5.32
70.7462	4.61	5.07	5.34
70.8909	4.29	4.75	5.01
70.9239	4.26	4.71	4.97
71.0034	4.11	4.55	4.81
71.2551	4.3	4.73	4.97
71.5027	2.83	3.24	3.48
71.7536	3.78	4.18	4.41
71.9985	3.43	3.82	4.04
72.253	3.09	3.45	3.68
72.5009	4.76	5.11	5.31

Table A-6: 100-year Water Surface Profile: Sacramento River

Rvr Sta (mi)	Levee Elevation (ft) West	Levee Elevation (ft) East	WSEL (ft) Existing	WSEL (ft) 900-Ft	WSEL (ft) 1,800-Ft
40.0	30.28	30.67	24.98	24.4	24.08
40.3	31.26	29.89	25.2	24.6	24.28
40.5	31.94	30.58	25.37	24.77	24.44
40.8	31.42	28.94	25.52	24.91	24.57
41.0	32.13	28.21	25.7	25.08	24.74
41.2	30.78	29.49	25.84	25.21	24.87
41.5	30.11	29.8	26	25.37	25.02
41.7	31.72	30.07	26.1	25.46	25.11
42.0	34.19	31.04	26.27	25.63	25.28
42.3	32.34	32.67	26.44	25.79	25.43
42.5	31.8	30.77	26.53	25.88	25.52
42.7	31.74	32.22	26.61	25.95	25.59
43.0	33.05	32.98	26.63	25.98	25.62
43.2	31.98	33.04	26.8	26.14	25.78
43.5	32.08	32.69	26.94	26.27	25.9
43.7	32.78	33.27	27.22	26.54	26.15
44.0	32.56	32.88	27.27	26.59	26.21
44.3	32.3	33.1	27.3	26.62	26.24
44.5	32.12	32.79	27.42	26.73	26.35
44.7	33.33	34.46	27.61	26.91	26.52
45.0	34.21	34.79	27.72	27.02	26.63
45.2	34.54	34.46	27.89	27.18	26.78
45.5	35.46	35.02	28.08	27.36	26.96
45.8	31.6	34.71	28.21	27.48	27.08
46.0	32.28	35.37	28.31	27.58	27.17
46.3	32.29	34.73	28.42	27.69	27.27
46.4	34.58	34.8	28.54	27.8	27.38
46.4	34.82	34.53	28.54	27.8	27.38
46.4	34.98	34.51	28.56	27.82	27.41
46.4	34.92	34.48	28.56	27.82	27.41
46.5	33.09	33.52	28.59	27.85	27.43
46.8	32.87	34.33	28.61	27.87	27.45
47.0	33.75	34.13	28.64	27.9	27.48

Rvr Sta (mi)	Levee Elevation (ft) West	Levee Elevation (ft) East	WSEL (ft) Existing	WSEL (ft) 900-Ft	WSEL (ft) 1,800-Ft
47.3	32.64	34.58	28.77	28.02	27.6
47.5	35.72	35.52	29.03	28.27	27.84
47.8	33.93	35.27	29.16	28.39	27.96
48.0	33.11	34.42	29.29	28.51	28.08
48.3	34.38	34.69	29.35	28.57	28.14
48.5	34.85	32.92	29.43	28.65	28.22
48.8	34.06	33.71	29.53	28.75	28.31
49.0	34.69	34.06	29.59	28.81	28.37
49.3	34.99	34.16	29.76	28.97	28.53
49.5	35.69	35.02	29.92	29.11	28.67
49.8	35.66	36.71	30.04	29.23	28.78
50.0	34.66	36.27	30.23	29.41	28.96
50.2	34.55	33.87	30.3	29.48	29.03
50.5	35.05	33.8	30.38	29.56	29.11
50.8	35.29	35.2	30.43	29.61	29.16
51.0	35.97	36.48	30.43	29.62	29.17
51.2	36.04	38.73	30.56	29.74	29.29
51.5	34.92	38.41	30.77	29.94	29.48
51.8	34.59	39.64	30.89	30.06	29.59
52.0	36.44	41.27	30.9	30.07	29.61
52.3	36.55	40.53	31	30.17	29.7
52.5	36.18	38.62	31.3	30.45	29.97
52.7	35.86	40.03	31.44	30.58	30.1
53.0	36.82	40.66	31.51	30.65	30.17
53.2	36.44	40.76	31.56	30.71	30.22
53.5	36.75	39.9	31.56	30.71	30.23
53.8	34.68	39.4	31.71	30.86	30.37
54.0	37.06	41.87	31.83	30.98	30.49
54.3	37.06	41.05	32.12	31.25	30.75
54.5	40.06	41.26	32.17	31.3	30.8
54.7	40.58	39.21	32.32	31.44	30.94
55.0	39.14	39.33	32.48	31.61	31.1
55.2	39.28	40.6	32.46	31.58	31.08

Rvr Sta (mi)	Levee Elevation (ft) West	Levee Elevation (ft) East	WSEL (ft) Existing	WSEL (ft) 900-Ft	WSEL (ft) 1,800-Ft
55.5	38.82	39.47	32.53	31.65	31.15
55.7	49.39	38.5	32.71	31.83	31.32
56.0	40.14	41.02	32.68	31.8	31.3
56.3	45.08	40.43	32.8	31.92	31.42
56.5	44.28	41.25	32.96	32.08	31.57
56.8	44.4	41.74	33.09	32.2	31.69
57.0	44.23	41.74	33.19	32.3	31.79
57.3	44.2	40.89	33.28	32.39	31.88
57.5	44.07	41.28	33.34	32.44	31.93
57.6		42.54	33.45	32.55	31.99
57.9			33.59	32.67	32.1
58.0			33.62	32.72	32.22
58.2			33.76	32.85	32.35
58.5	37.9	39.54	33.85	32.94	32.44
58.5	37.87	38.43	33.85	32.94	32.44
58.5	38.13	39.64	33.85	32.94	32.44
58.5	38.13	39.65	33.85	32.94	32.44
58.5	38.13	39.65	33.85	32.94	32.44
58.5	38.13	39.66	33.85	32.94	32.44
58.5	38.13	39.7	33.86	32.96	32.45
58.5	38.13	39.71	33.86	32.95	32.45
58.8	37.99	40.46	33.98	33.07	32.56
59.0	36.06	40.34	34.1	33.19	32.68
59.3	36	40.07	34.23	33.31	32.8
59.3	36.81	40.29	34.23	33.31	32.8
59.3	36.74	40.68	34.22	33.31	32.8
59.3	36.26		34.26	33.35	32.84
59.3	35.35		34.26	33.35	32.84
59.5	33.58	40.66	34.22	33.31	32.8
59.7	39.09		34.38	33.46	32.94
59.7	39.09		34.38	33.46	32.94
59.7	39.09		34.38	33.46	32.94
59.7			34.39	33.47	32.96

Rvr Sta (mi)	Levee Elevation (ft) West	Levee Elevation (ft) East	WSEL (ft) Existing	WSEL (ft) 900-Ft	WSEL (ft) 1,800-Ft
59.7	39	38.41	34.39	33.47	32.96
59.7	38.85	38.98	34.39	33.47	32.96
59.8	37.82	40.41	34.36	33.44	32.92
60.0	61.62	42.4	34.54	33.61	33.09
60.3	44.02	42.93	34.69	33.71	33.16
60.4	43.74	42.89	34.69	33.78	33.26
60.4	43.75	42.87	34.69	33.78	33.26
60.6		35.67	34.69	33.78	33.26
60.6		35.68	34.69	33.78	33.26
60.8		45.52	34.71	33.78	33.27
60.9		45.79	34.71	33.79	33.27
60.9	29.4	45.79	34.71	33.79	33.27
61.0	42.3	44.02	34.71	33.79	33.27
61.0	42.3	44.02	34.71	33.79	33.27
61.3	41.56	43.39	34.71	33.79	33.27
61.5	42.44	43.2	34.71	33.79	33.27
61.7	41.75	43.07	34.71	33.79	33.27
62.0	41.56	44.03	34.71	33.79	33.27
62.3	41.56	41.32	34.71	33.79	33.27
62.5	42.06	39.43	34.71	33.79	33.27
62.8	42.2	43.16	34.71	33.79	33.27
62.9	41.63	41.24	34.71	33.79	33.27
62.9	41.63	41.28	34.71	33.79	33.27
63.0	41.76	41.41	34.71	33.79	33.27
63.0	41.79	41.44	34.71	33.79	33.27
63.0	41.85	41.46	34.71	33.79	33.27
63.0	41.86	41.43	34.71	33.79	33.27
63.0	41.87	41.33	34.71	33.79	33.27
63.0	41.87	41.31	34.71	33.79	33.27
63.0	41.84	41.18	34.71	33.79	33.27
63.3	41.25	40.43	34.71	33.79	33.27
63.3	41.23	40.43	34.71	33.79	33.27
63.3	41.2	40.43	34.71	33.79	33.27

Rvr Sta (mi)	Levee Elevation (ft) West	Levee Elevation (ft) East	WSEL (ft) Existing	WSEL (ft) 900-Ft	WSEL (ft) 1,800-Ft
63.3	41.18	40.43	34.71	33.79	33.27
63.3	41.15	40.43	34.71	33.79	33.27
63.3	41.13	40.43	34.71	33.79	33.27
63.4	41.1	40.43	34.71	33.79	33.27
63.4	41.08	40.43	34.71	33.79	33.27
63.4	41.06	40.43	34.71	33.79	33.27
63.4	41.03	40.43	34.71	33.79	33.27
63.4	41.01	40.43	34.71	33.79	33.27
63.4	40.98	40.43	34.71	33.79	33.27
63.5	40.96	40.43	34.71	33.79	33.27
63.5	40.93	40.43	34.71	33.79	33.27
63.5	40.91	40.43	34.7	33.79	33.27
63.5	40.89	40.44	34.69	33.78	33.27
63.5	40.86	40.45	34.67	33.77	33.26
63.6	40.84	40.46	34.65	33.76	33.25
63.6	40.82	40.47	34.63	33.73	33.23
63.6	40.79	40.48	34.59	33.72	33.22
63.6	40.77	40.49	34.56	33.69	33.19
63.6	40.74	40.51	34.54	33.66	33.17
63.6	40.72	40.52	34.49	33.63	33.14
63.7	40.7	40.53	34.45	33.6	33.12
63.7	40.67	40.54	34.4	33.56	33.09
63.7	40.65	40.55	34.37	33.53	33.07
63.7	40.63	40.56	34.34	33.49	33.03
63.7	40.6	40.57	34.28	33.46	33.01
63.8	40.58	40.58	34.25	33.42	32.98
63.8	40.57	40.41	34.21	33.37	32.94
63.8	40.57	40.25	34.15	33.34	32.91
63.8	40.56	40.08	34.12	33.29	32.87
63.8	40.56	39.92	34.07	33.24	32.83
63.8	40.55	39.75	34.04	33.21	32.79
63.9	40.55	39.59	34.05	33.17	32.77
63.9	40.54	39.42	34.06	33.13	32.74

Rvr Sta (mi)	Levee Elevation (ft) West	Levee Elevation (ft) East	WSEL (ft) Existing	WSEL (ft) 900-Ft	WSEL (ft) 1,800-Ft
63.9	40.54	39.26	34.07	33.12	32.71
63.9	40.53	39.09	34.08	33.1	32.71
63.9	40.53	38.93	34.1	33.09	32.69
64.0	40.52	38.76	34.11	33.09	32.69
64.0	40.52	38.6	34.12	33.09	32.69
64.0	40.51	38.43	34.14	33.09	32.69
64.3	41.04	39.16	34.37	33.37	32.75
64.5	41.15	39.43	34.46	33.48	32.88
64.8	41.33	40.82	34.59	33.62	33.03
65.0	41.41	40.39	34.76	33.82	33.25
65.3	41.68	40.81	34.96	34.05	33.5
65.5	41.86	41.49	35.19	34.31	33.79
65.8	42.58	41.63	35.26	34.4	33.89
66.0	42.39	42.02	35.36	34.52	34.02
66.3	41.53	42.55	35.49	34.66	34.17
66.5	41.26	43.61	35.81	35.03	34.56
66.8	41.67	40.62	35.98	35.23	34.78
67.0	41.95	42.01	36.3	35.58	35.16
67.3	41.32	41.15	36.37	35.66	35.24
67.5	41.32	39.95	36.5	35.81	35.41
67.8	41.2	41.02	36.69	36.02	35.63
68.0	41.25	40.42	36.96	36.33	35.96
68.2	41.19	40.39	37.17	36.56	36.2
68.5	40.32	41.39	37.42	36.84	36.5
68.7	40.68	41.48	37.51	36.94	36.61
69.0	41.1	41.4	37.71	37.16	36.84
69.3	40.65	42.62	37.84	37.31	36.99
69.5	41.13	42.52	37.92	37.39	37.09
69.8	41.16	42.11	38.02	37.51	37.21
70.0	41.21	42.27	38.2	37.7	37.42
70.2	41.38	42	38.26	37.78	37.5
70.5	41.76	42.92	38.35	37.87	37.6
70.7	42.13	43.06	38.45	37.99	37.72

Rvr Sta (mi)	Levee Elevation (ft) West	Levee Elevation (ft) East	WSEL (ft) Existing	WSEL (ft) 900-Ft	WSEL (ft) 1,800-Ft
70.9	41.56	42.82	38.53	38.07	37.81
70.9	41.62	42.82	38.56	38.11	37.85
71.0	41.64	42.77	38.66	38.22	37.96
71.3	42.53	43.19	38.89	38.46	38.22
71.5	42.1	41.88	39.05	38.64	38.4
71.8	42.12	42.94	39.16	38.76	38.53
72.0	41.33	42.64	39.21	38.82	38.6
72.3	41.32	42.61	39.52	39.16	38.95
72.5	42.76	44.45	39.69	39.34	39.14
72.8	42.23	43.78	39.91	39.58	39.39
73.0	43.08	43.99	40.02	39.7	39.51
73.3	43.02	41.53	40.21	39.9	39.73
73.5	42.36	43.4	40.32	40.02	39.85
73.7	43.16	43.27	40.44	40.15	39.99
74.0	42.36	41.89	40.55	40.27	40.12
74.3	43.19	41.1	40.68	40.41	40.26
74.5	43.49	41.21	40.73	40.47	40.33
74.8	44.19	43.03	41.02	40.78	40.64
75.0	43.42	41.76	41.15	40.92	40.8
75.3	42.36	41.57	41.25	41.08	40.9
75.5	42.4	42.15	41.41	41.19	41.08
75.8	42.66	41.38	41.57	41.37	41.25
76.0	43.08	43.35	41.67	41.48	41.37
76.3	42.89	42.09	41.85	41.67	41.57
76.5	43.26	41.22	41.98	41.81	41.72
76.7	43.58	43.01	42.06	41.89	41.8
77.0	43.42	42.09	42.19	42.04	41.95
77.3	43.97	42.46	42.34	42.2	42.12
77.5	44.06	43.84	42.39	42.25	42.17
77.8	44.41	45.72	42.63	42.51	42.44
78.0	44.93	45.27	42.84	42.73	42.67
78.3	44.87	43.89	42.92	42.81	42.75
78.5	45.05	45.42	43.04	42.84	42.88

Rvr Sta (mi)	Levee Elevation (ft) West	Levee Elevation (ft) East	WSEL (ft) Existing	WSEL (ft) 900-Ft	WSEL (ft) 1,800-Ft
78.8	44.06	43.48	43.2	43.11	43.06
79.0	44.21	48.44	43.34	43.26	43.21
79.2	46.22	48	43.49	43.42	43.38
79.2	46.24	48	43.49	43.42	43.38
79.3	46.16	45.4	43.49	43.42	43.38
79.3	46.16	45.4	43.49	43.42	43.38
79.5	47.02	46.26	43.63	43.56	43.52
79.8	47.86	47.37	43.73	43.67	43.63
80.0	46.35	45.6	43.78	43.72	43.69
80.0	46.28	45.6	43.78	43.72	43.69
80.4		46.63	43.78	43.72	43.69
80.5		46.82	43.83	43.78	43.75
80.8		47.78	43.86	43.81	43.78
81.3		45.56	43.86	43.81	43.79
81.5		44.96	43.86	43.82	43.79
81.8		46.19	43.87	43.82	43.79
82.0		47.55	43.72	43.67	43.65
82.3			43.72	43.67	43.65
82.5			43.76	43.71	43.68
82.7			43.76	43.72	43.69
83.0			43.77	43.73	43.7
83.3			43.7	43.65	43.63
83.5			43.78	43.74	43.71
83.8	47.35		43.61	43.57	43.54
84.0	48.4		43.63	43.59	43.56
84.3	47.81		43.63	43.59	43.57
84.5	47.17		43.64	43.6	43.58
84.8	47.21	47.94	43.64	43.6	43.57
85.0	47.1	48.12	43.62	43.57	43.55
85.3	45.47	47.62	43.67	43.62	43.6
85.5	45.06	47.81	43.69	43.64	43.62
85.8	45.08	47.21	43.73	43.69	43.66
86.0	45.16	45.7	43.76	43.72	43.7

Rvr Sta (mi)	Levee Elevation (ft) West	Levee Elevation (ft) East	WSEL (ft) Existing	WSEL (ft) 900-Ft	WSEL (ft) 1,800-Ft
86.3	44.4	43.7	43.8	43.76	43.74
86.5	44.71	44.94	43.82	43.79	43.77
86.8	44.57	44.28	43.86	43.83	43.81
87.0	45.16	45.16	43.9	43.87	43.85
87.3	45.55	44.73	43.96	43.93	43.91
87.5	45.23	47.31	44.04	44.01	43.99
87.8	46.23	46.41	44	43.97	43.95
88.0	46.7	50	44.16	44.13	44.12
88.3	46.37	45.65	44.18	44.15	44.13
88.5	46.38	44.46	44.17	44.14	44.12
88.8	46.69	45.12	44.16	44.13	44.11
89.0	46.98	50.91	44.23	44.2	44.18
89.3	46.08	46.6	44.25	44.22	44.2
89.5	48.13	45.51	44.31	44.28	44.27
89.8	47.6	46.73	44.34	44.31	44.29
90.0	47.47	46.13	44.42	44.39	44.37
90.2	46.24	46.41	44.4	44.37	44.36
90.2	46.26	46.3	44.4	44.37	44.35
90.2	46.7	45.11	44.47	44.44	44.43
90.2	46.81	44.98	44.48	44.45	44.44
90.3	47.4	45.48	44.52	44.49	44.47
90.3	47.6	45.69	44.55	44.53	44.51
90.4	47.35	47.37	44.56	44.54	44.52
90.5	47.58	47	44.6	44.57	44.55
90.8	47.6	46.57	44.63	44.6	44.58
91.0	47.79	47	44.7	44.67	44.66

Table A-7: 100-year Water Surface Profile: Yolo Bypass

Rvr Sta (mi)	Levee Elevation (ft) West	Levee Elevation (ft) East	WSEL (ft) Existing	WSEL (ft) 900-Ft	WSEL (ft) 1,800-Ft
15.208	20.84	10.54	11.22	11.21	11.2
15.2081	20.84	10.54	11.22	11.21	11.2
15.457	20.09	11.15	11.28	11.27	11.26
15.708	21.92	12.87	12.73	12.73	12.73
15.957	21.78	11.39	14.3	14.3	14.31
16.207	24.43	11.3	15.09	15.1	15.11
16.461	21.18	11.42	15.66	15.67	15.68
16.707	22.34	11.6	15.91	15.93	15.93
16.956	20.87	11.87	16.07	16.08	16.09
17.208	20.72	11.75	16.26	16.27	16.28
17.457	21.82	11.89	16.63	16.64	16.65
17.708	21.9	11.65	17.11	17.13	17.14
17.958	22.74	13.14	17.4	17.42	17.43
18.208	22.62	14.22	18.18	18.22	18.24
18.4738	20.96	14.97	19.58	19.62	19.64
18.474	20.96	14.97	19.58	19.62	19.64
18.767	11.71	24.53	19.58	19.62	19.64
18.7671	11.71	24.53	19.58	19.62	19.64
18.974	8.44	23.38	19.62	19.66	19.68
19.161	11.32	22.66	19.67	19.72	19.73
19.3478	10.49	24.46	19.75	19.79	19.81
19.348	10.49	24.46	19.75	19.79	19.81
19.534	10.07	20.55	19.75	19.79	19.81
19.5341	10.07	20.55	19.75	19.79	19.81
19.721	9.39	21.63	19.77	19.82	19.84
19.907	9.87	19.76	19.82	19.87	19.89
20.093	15.02	20.93	19.86	19.91	19.93
20.28	17.99	21.5	19.91	19.95	19.97
20.466	19.02	21.47	19.96	20	20.02
20.6528	18.73	19.58	20.01	20.06	20.08
20.653	18.73	19.58	20.01	20.06	20.08
20.839	18.2	20.03	20.01	20.06	20.08
20.8391	18.2	20	20.01	20.06	20.08

Rvr Sta (mi)	Levee Elevation (ft) West	Levee Elevation (ft) East	WSEL (ft) Existing	WSEL (ft) 900-Ft	WSEL (ft) 1,800-Ft
21.025	16.74	22.79	20.04	20.09	20.11
21.212	19.34	21.91	20.13	20.17	20.19
21.398	19.11	23.55	20.18	20.23	20.25
21.585	19.46	18.82	20.25	20.3	20.33
21.771	18.98	22.04	20.32	20.37	20.39
21.958	20.02	22.59	20.39	20.44	20.46
22.144	20.64	23.11	20.45	20.5	20.52
22.331	20.22	23.17	20.53	20.58	20.6
22.518	28.72	23.96	20.62	20.67	20.69
22.705	28.21	23.56	20.71	20.77	20.79
22.892	27.5	24.24	20.81	20.86	20.89
23.079	27.7	23.7	20.9	20.95	20.98
23.266	28.79	24.23	20.98	21.04	21.07
23.453	28.59	24.57	21.07	21.13	21.16
23.64	26.73	25.01	21.15	21.21	21.24
23.826	26.69	25.87	21.23	21.29	21.32
24.013	27.73	25.88	21.34	21.4	21.43
24.2	27.52	25.1	21.43	21.5	21.53
24.387	26.94	25.47	21.53	21.6	21.63
24.574	28.1	26.22	21.62	21.69	21.72
24.761	28.07	25.65	21.73	21.79	21.83
24.948	28.05	24.71	21.83	21.9	21.93
25.135	28.74	25	21.94	22.01	22.04
25.322	27.79	25.35	22.04	22.11	22.15
25.508	27.9	26.47	22.14	22.21	22.25
25.695	27.65	27.45	22.25	22.32	22.36
25.883	27.8	27.29	22.38	22.45	22.49
26.071	27.82	27.32	22.48	22.56	22.6
26.259	27.74	27.27	22.6	22.67	22.71
26.447	25.65	27.61	22.72	22.79	22.83
26.635	27.95	27.55	22.82	22.89	22.93
26.823	27.9	27.15	22.9	22.98	23.02
27.011	28.58	27.26	22.99	23.07	23.11

Rvr Sta (mi)	Levee Elevation (ft) West	Levee Elevation (ft) East	WSEL (ft) Existing	WSEL (ft) 900-Ft	WSEL (ft) 1,800-Ft
27.199	27.83	27.48	23.07	23.15	23.19
27.387	29.21	28.38	23.14	23.22	23.26
27.575	29.12	28.47	23.22	23.3	23.34
27.763	29.02	28.41	23.31	23.39	23.43
27.951	30.31	28.78	23.41	23.49	23.53
28.139	29.71	28.53	23.51	23.59	23.63
28.327	28.73	28.41	23.63	23.71	23.75
28.515	28.86	26.76	23.77	23.86	23.9
28.703	29.01	27.36	23.92	24	24.05
28.891	29.68	28.32	24.09	24.18	24.22
29.079	29.43	27.08	24.41	24.49	24.54
29.267	31.3	28.28	24.69	24.77	24.81
29.455	29.2	25.67	24.95	25.03	25.07
29.643	28.62	28.56	25.14	25.22	25.26
29.835	28.89	28.81	25.27	25.35	25.39
30.026	29.3	28.96	25.39	25.47	25.52
30.217	29.06	29.71	25.52	25.6	25.65
30.408	28.16	25.73	25.64	25.73	25.77
30.599	28.96	26.89	25.76	25.84	25.89
30.79	29.08	24.83	25.87	25.96	26
30.981	29.14	29.27	25.97	26.06	26.1
31.172	31.89	29.71	26.1	26.18	26.23
31.363	32.05	29.74	26.32	26.41	26.45
31.554	32.52	30.89	26.53	26.62	26.66
31.745	33.08	32.01	26.7	26.78	26.83
31.936	32.83	33.28	26.83	26.92	26.97
32.131	33.13	33.43	26.97	27.05	27.1
32.325	33.7	33.7	27.1	27.19	27.23
32.52	32.84	32.73	27.25	27.34	27.38
32.714	30.44	31.02	27.4	27.48	27.53
32.909	31.09	30.18	27.49	27.58	27.62
33.103	31	30.52	27.58	27.67	27.71
33.298	31.81	26.21	27.67	27.76	27.8

Rvr Sta	Levee Elevation (ft)		WSEL (ft)		
(mi)	West	East	Existing	900-Ft	1,800-Ft
33.492	32	31.41	27.75	27.84	27.89
33.687	32.16	32.67	27.84	27.92	27.97
33.882	31.76	34.21	27.91	28	28.04
34.079	31.72	32.47	27.98	28.06	28.11
34.276	32.01	33.77	28.04	28.13	28.17
34.474	31.58	32.88	28.1	28.19	28.24
34.671	31.7	34.31	28.17	28.25	28.3
34.869	31.09	32.82	28.24	28.32	28.37
35.066	31.22	32.14	28.31	28.4	28.44
35.268	31.52	34.18	28.39	28.48	28.53
35.47	30.93	31.61	28.48	28.57	28.62
35.672	29.75	32.11	28.57	28.66	28.71
35.874	31.79	33.73	28.67	28.76	28.8
36.077	32.16	33.03	28.77	28.85	28.9
36.279	32.06	33.95	28.85	28.94	28.99
36.487	32.47	32.47	28.92	29.01	29.06
36.695	32.09	33.4	29	29.09	29.14
36.904	31.75	33.91	29.07	29.16	29.21
37.112	32.95	33.54	29.14	29.23	29.28
37.32	33.14	37.52	29.21	29.3	29.35
37.529	32.36	38.94	29.28	29.37	29.42
37.727	33.22	34.59	29.36	29.45	29.5
37.926	32.58	38.72	29.46	29.55	29.6
38.125	33.05	33.24	29.58	29.67	29.72
38.324	33.42	38.28	29.7	29.78	29.83
38.5218	33.53		29.85	29.94	29.98
38.522	33.53		29.86	29.95	29.99
38.656	33.02		29.86	29.95	29.99
38.6561	33.02		29.86	29.94	29.99
38.813	32.65		29.96	30.04	30.09
38.938	33.68		30.04	30.13	30.17
39.084	33.51		30.09	30.17	30.22
39.271	33.78	33.22	30.21	30.29	30.34

Rvr Sta	Levee Elevation (ft)		WSEL (ft)		
(mi)	West	East	Existing	900-Ft	1,800-Ft
39.458	33.62	32.58	30.31	30.39	30.44
39.645	33.72	32.3	30.39	30.48	30.53
39.831	33.69	34.09	30.47	30.56	30.61
40.018	33.1	33.24	30.55	30.64	30.68
40.204	33.75	33.41	30.6	30.7	30.74
40.39	33.89	33.37	30.67	30.76	30.81
40.577	34.44	32.32	30.73	30.82	30.87
40.763	35.57	33.04	30.79	30.88	30.93
40.95	35.64	32.83	30.85	30.94	30.99
41.136	34.76	33.78	30.9	31	31.05
41.322	33.27	33.68	30.95	31.05	31.1
41.509	33.61	33.65	31.01	31.1	31.15
41.695	33.34	33.51	31.06	31.15	31.2
41.882	33.45	33.71	31.11	31.2	31.25
42.11	34.04	33.72	31.16	31.26	31.31
42.342	33.29	33.53	31.21	31.31	31.36
42.51	33.47	35.71	31.38	31.48	31.53
42.62	33.6	34.9	31.41	31.51	31.56
42.6893	33.6	34.61	31.42	31.52	31.57
42.7586	33.6	34.33	31.58	31.68	31.73
42.828	33.6	34.04	31.59	31.69	31.74
42.969	33.6	36.81	31.59	31.69	31.74
42.9691	33.6	36.81	31.59	31.69	31.74
43.089	33.6	36.64	31.6	31.7	31.76
43.375	33.55	35.6	31.65	31.75	31.81
43.582	33.55	37.15	31.69	31.79	31.84
43.751	33.65	36.54	31.72	31.82	31.88
43.9488	33.88	36.5	31.76	31.86	31.92
43.949	33.88	36.5	31.76	31.86	31.92
44.13	34.98	34.98	31.76	31.86	31.92
44.1301	34.98	34.98	31.76	31.86	31.92
44.498	32.32	35.74	31.82	31.92	31.97
44.81	33.31	34.52	31.88	31.97	32.02

Rvr Sta	Levee Elevation (ft)		WSEL (ft)		
(mi)	West	East	Existing	900-Ft	1,800-Ft
45.059	33.15	34.82	31.93	32.03	32.07
45.37	33.44	34.9	32.01	32.1	32.14
45.598	33.81	35.96	32.07	32.16	32.21
45.829	34.36	35.96	32.14	32.23	32.27
46.245	33.44	35.91	32.28	32.36	32.4
46.452	33.32	37.22	32.35	32.43	32.47
46.658	34.3	36.21	32.43	32.51	32.55
46.857	35.14	36.42	32.51	32.59	32.62
47.076	34.94	34.86	32.6	32.68	32.72
47.279	35.68	36.62	32.69	32.76	32.8
47.488	36.78	37.44	32.77	32.84	32.88
47.711	35.88	37.44	32.87	32.94	32.97
47.92	35.61	36.9	32.96	33.02	33.05
48.124	36.29	35.86	33.03	33.1	33.13
48.327	35.47	33.82	33.11	33.18	33.2
48.558	36.06	35.43	33.19	33.25	33.28
48.84	37.7	35.66	33.3	33.36	33.38
49.105	36.89	38.08	33.44	33.5	33.52
49.385	37.94	37.68	33.65	33.7	33.72
49.653	37.78	38.4	33.86	33.91	33.93
50.084	38.63	38.97	34.22	34.26	34.28
50.2895	36.69	38.85	34.46	34.49	34.51
50.495	34.75	38.74	34.64	34.67	34.68
50.496	34.75	38.74	34.64	34.67	34.68
50.563	36.82	42.38	34.71	34.73	34.75
50.63	38.9	46.01	35.06	35.09	35.09
50.697	40.97	49.65	35.13	35.15	35.15
50.911	39.66	49.98	35.31	35.32	35.33
51.097	39.99	34.89	35.49	35.5	35.51
51.284	39.85	50.55	35.68	35.69	35.69
51.47	40.02	48.69	35.82	35.83	35.83
51.657	41.58	49.4	35.98	35.98	35.98
51.844	40.68	49.4	36.13	36.13	36.13

Rvr Sta	Levee Elevation (ft)		WSEL (ft)		
(mi)	West	East	Existing	900-Ft	1,800-Ft
52.043	40.98	50.68	36.28	36.28	36.28
52.243	42.02	42.4	36.39	36.39	36.38
52.442	41.13	42.14	36.48	36.48	36.48
52.642	40.67	41.25	36.58	36.58	36.58
53.041	40.34	41.78	36.79	36.78	36.78
53.24	41.31	42.4	36.88	36.88	36.87
53.44	42.88	42.04	36.98	36.97	36.97
53.533	42.64	41.17	37.02	37.02	37.01
53.6258	42.45	40.36	37.07	37.06	37.05
53.626	42.4	40.31	37.07	37.06	37.05
53.813	42.63	40.9	37.07	37.06	37.05
53.8132	42.63	40.9	37.09	37.09	37.08
53.999	43.02	42.44	37.16	37.15	37.15
54.186	42.3	42.38	37.27	37.26	37.25
54.372	42.63	43.59	37.39	37.38	37.37
54.558	44.36	44.1	37.52	37.51	37.5
54.745	43.51	44.5	37.65	37.64	37.63
54.931	43.44	44.33	37.8	37.79	37.78
55.118	43.84	44.2	37.95	37.94	37.92
55.304	43.36	44.9	38.16	38.14	38.13
55.49	44.54	44.81	38.4	38.38	38.37
55.677	44.74	43.2	38.68	38.66	38.65
55.863	44.21	44.9	39	38.97	38.96
56.05	45.06	45.58	39.27	39.24	39.23
56.236	44.8	45.07	39.61	39.59	39.57
56.629	44.95	45.07	40.37	40.34	40.32

Table A-8: 100-year Water Surface Profile: Sacramento Bypass

Rvr Sta	Levee Elevation (ft)		WSEL (ft)		
(mi)	South	North	Existing	900-Ft	1,800-Ft
0		36.03	31.76	31.86	31.92
0.000183		36.03	31.76	31.86	31.92
0.18	35.06	34.28	31.8	31.89	31.94
0.37	34.82	35.17	32.01	31.98	31.99
0.56	35.39	36.27	32.2	32.08	32.05
0.75	37.91	36.63	32.37	32.17	32.1
0.93	36.06	37.14	32.59	32.26	32.16
1.12	37.36	38.35	32.82	32.38	32.22
1.3	36.87	38.04	33.01	32.49	32.3
1.49	36.82	38.4	33.28	32.66	32.4
1.67981	39.66	40.9	33.65	32.87	32.54
1.68	39.66	40.9	34.28	33.27	32.81

Table A-9: 100-year Water Surface Profile: American River

Rvr Sta	Levee Elevation (ft)		WSEL (ft)		
(mi)	South	North	Existing	900-Ft	1,800-Ft
0.12	43.37		34.69	33.78	33.26
0.18	45.16		34.81	33.9	33.39
0.18	45.52		34.81	33.9	33.39
0.19	44.71		34.86	33.96	33.46
0.19	44.39		34.87	33.96	33.46
0.23	53.56		34.86	33.96	33.46
0.26	55.35		34.95	34.07	33.57
0.26	52.9		34.96	34.07	33.58
0.27	53.09		34.98	34.1	33.61
0.27	58.06		34.98	34.1	33.61
0.27	55.65		34.98	34.09	33.6
0.29	55.84		35.01	34.13	33.64
0.29	58.18		35	34.12	33.64
0.36	47.09		35.04	34.17	33.68
0.43	41.48		35.12	34.26	33.78
0.50	41.42		35.18	34.33	33.86
0.58	41.45		35.24	34.4	33.94
0.67	41.48		35.3	34.48	34
0.75	41.57		35.35	34.53	34.07
0.83	41.48		35.42	34.6	34.15
0.92	41.29		35.48	34.67	34.22
1.00	41.1		35.54	34.74	34.3
1.08	41.05		35.61	34.81	34.38
1.17	41.2		35.68	34.89	34.46
1.25	41.35		35.75	34.97	34.54
1.33	41.42		35.76	34.99	34.57
1.42	41.35		35.81	35.04	34.62
1.50	42.09		35.86	35.09	34.68
1.58	41.47		35.88	35.12	34.71
1.66	41.65		35.88	35.12	34.71
1.74	41.83		35.9	35.14	34.73
1.75	41.83		35.89	35.13	34.72
1.75	41.85		35.89	35.14	34.73

Rvr Sta	Levee Elevation (ft)		WSEL (ft)		
(mi)	South	North	Existing	900-Ft	1,800-Ft
1.75	41.85		35.89	35.13	34.72
1.82	42.07		35.95	35.2	34.79
1.88	43.3	43.06	36.01	35.27	34.87
1.95	44.55	43.8	36.07	35.33	34.94
1.98	47.38	43.47	36.13	35.41	35.01
1.99	45.23	43.49	36.23	35.52	35.13
2.01	49.1	43.53	36.29	35.58	35.2
2.02	49.1	43.55	36.3	35.59	35.21
2.04	48	43.78	36.32	35.62	35.24
2.07	45.85	43.66	36.35	35.65	35.27
2.07	45.75	43.66	36.36	35.66	35.28
2.07	45.72	43.67	36.39	35.68	35.3
2.07	45.63	43.67	36.39	35.68	35.3
2.16	45.65	43.87	36.5	35.81	35.43
2.17	45.7	43.87	36.51	35.83	35.44
2.17	45.65	43.87	36.54	35.85	35.47
2.17	45.7	43.88	36.54	35.85	35.48
2.25	44.7	44.05	36.59	35.9	35.53
2.27	45	44.09	36.63	35.94	35.57
2.27	44.7	44.1	36.57	35.88	35.51
2.27	44.7	44.1	36.6	35.91	35.54
2.27	44.7	44.1	36.6	35.9	35.53
2.35	44.71	44.27	36.77	36.09	35.72
2.42	44.99	44.43	36.83	36.16	35.79
2.50	45	44.57	36.88	36.21	35.85
2.58	45.01	44.73	36.95	36.29	35.93
2.67	45.02	44.9	37	36.34	35.99
2.75	46.48	45.05	37.06	36.4	36.05
2.83	57.25	45.2	37.08	36.43	36.08
2.92	67.9	45.32	37.11	36.46	36.11
3.00	78.56	45.42	37.15	36.5	36.16
3.08	80.01	45.51	37.17	36.53	36.19
3.17	81.46	45.64	37.2	36.57	36.23

Rvr Sta	Levee Elevation (ft)		WSEL (ft)		
(mi)	South	North	Existing	900-Ft	1,800-Ft
3.25	82.91	45.83	37.24	36.6	36.26
3.33	77.77	46.02	37.34	36.72	36.38
3.42	72.64	46.21	37.45	36.84	36.51
3.50	67.5	46.35	37.58	36.96	36.64
3.57	55.94	46.34	37.67	37.08	36.77
3.64	48.55	47.28	37.78	37.21	36.9
3.64	48.9	47.28	37.78	37.21	36.9
3.65	48.54	47.1	37.79	37.22	36.91
3.65	48.54	47.1	37.8	37.22	36.92
3.70	48.61	46.68	37.74	37.16	36.86
3.75	48.78	46.7	37.85	37.28	36.98
3.82	48.69	49.88	37.94	37.36	37.07
3.89	50.3	50.9	38.04	37.49	37.2
3.89	49.3	56.97	38.04	37.49	37.19
3.91	53.99	50.39	38.11	37.57	37.28
3.91	55.96	56.96	38.12	37.58	37.29
4.00	48.96	47.8	38.15	37.6	37.31
4.08	49.01	47.89	38.25	37.71	37.42
4.17	49.27	48.44	38.36	37.83	37.55
4.25	49.13	53.05	38.48	37.96	37.69
4.33	48.98	48.79	38.57	38.06	37.79
4.42	48.84	50.43	38.67	38.17	37.91
4.50	48.69	52.09	38.78	38.3	38.04
4.58	48.55	51.78	38.87	38.39	38.14
4.67	48.66	51.54	38.95	38.47	38.23
4.75	49.09	50.13	39.03	38.56	38.32
4.83	49.53	48.72	39.1	38.64	38.41
4.92	49.88	47.66	39.2	38.75	38.52
5.00	50.14	48.15	39.32	38.88	38.66
5.08	50.4	46.47	39.42	39	38.78
5.17	50.64	45.91	39.57	39.16	38.95
5.25	50.69	46.02	39.79	39.4	39.19
5.33	50.74	46.13	39.97	39.59	39.4

Rvr	Levee Elevation	WSEL (ft)	Rvr Sta	Levee Elevation	WSEL (ft)	Rvr Sta	Levee Elevation	WSEL (ft)

Sta	(ft)				
(mi)	South	North	Existing	900-Ft	1,800-Ft
5.42	50.79	46.24	40.16	39.79	39.61
5.50	51.25	46.62	40.36	40	39.82
5.58	50.99	46.61	40.45	40.1	39.92
5.67	51.1	46.79	40.54	40.2	40.03
5.75	51.47	46.59	40.64	40.3	40.13
5.83	51.85	46.36	40.6	40.26	40.09
5.92	51.86	46.44	40.48	40.14	39.96
6.00	51.64	46.66	40.32	39.97	39.79
6.08	51.45	46.88	40.56	40.23	40.06
6.17	51.33	47.1	40.98	40.67	40.51
6.25	51.51	47.5	41.52	41.24	41.1
6.32	51.67	47.48	41.99	41.75	41.62
6.39	51.84	47.65	42.38	42.15	42.03
6.46	52.4	47.81	42.69	42.48	42.37
6.46	53.87	47.82	42.71	42.49	42.38
6.47	53.48	47.84	42.77	42.56	42.45
6.47	52	47.99	42.78	42.56	42.45
6.50	52.06	47.9	42.82	42.6	42.5
6.58	52.24	48.09	43.01	42.81	42.7
6.67	52.54	48.39	43.22	43.02	42.92
6.75	52.99	48.83	43.42	43.23	43.13
6.82	53.35	49.17	43.59	43.4	43.3
6.88	53.6	50.74	43.77	43.59	43.49
6.95	53.38	53.16	43.98	43.8	43.71
6.95	53.38	53.21	44.01	43.83	43.74
6.95	53.4	53.4	44.06	43.88	43.79
6.95	53.37	52.32	44.03	43.85	43.76
7.00	53.22	50.11	44.24	44.06	43.97
7.08	52.93	50.42	44.66	44.49	44.41
7.17	52.88	50.75	45.01	44.85	44.77
7.25	52.88	51.05	45.27	45.12	45.04
7.33	53.21	51.37	45.32	45.17	45.09
7.42	53.8	51.68	45.42	45.27	45.19

	(ft)				
(mi)	South	North	Existing	900-Ft	1,800-Ft
7.50	54.4	52	45.56	45.41	45.34
7.56	55.48	52.21	45.72	45.57	45.5
7.61	58.62	54.5	45.87	45.72	45.65
7.61	59.33	56.58	45.93	45.79	45.72
7.62	59.33	56.58	45.95	45.81	45.74
7.62	59.5	56.69	45.89	45.74	45.67
7.62	59.5	56.7	45.88	45.74	45.67
7.62	54.63	52.51	45.91	45.77	45.7
7.63	54.6	52.54	45.93	45.79	45.72
7.63	58.35	55.9	45.87	45.73	45.65
7.69	54.3	54.22	46.01	45.87	45.8
7.75	54	53.06	46.13	45.99	45.92
7.83	53.58	53.44	46.34	46.21	46.14
7.92	53.65	53.72	46.49	46.36	46.29
8.00	54.22	54.16	46.62	46.49	46.42
8.08	54.02	54.26	46.66	46.53	46.46
8.17	54.21	54.53	46.68	46.55	46.48
8.25	54.7	54.83	46.67	46.54	46.48
8.33	54.62	55.07	46.74	46.61	46.55
8.42	54.9	55.34	46.82	46.69	46.63
8.50	55.2	55.61	46.92	46.8	46.73
8.58	55.49	55.89	47.07	46.95	46.88
8.67	55.76	55.94	47.24	47.12	47.06
8.75	56.03	56	47.44	47.32	47.27
8.83	56.09	56.06	47.69	47.57	47.51
8.92	56.07	56.15	47.95	47.84	47.78
9.00	56.05	56.25	48.15	48.04	47.99
9.07	57.97	57.48	48.09	47.98	47.93
9.13	60.39	60.25	47.96	47.85	47.8
9.13	59.19	60.08	47.94	47.83	47.78
9.15	55.97	56.63	47.93	47.82	47.76
9.15	58.97	60.04	47.95	47.84	47.79
9.20	55.95	58.72	48.05	47.94	47.88

	(ft)				
(mi)	South	North	Existing	900-Ft	1,800-Ft
9.25	57.23	57.63	48.13	48.03	47.97
9.33	56.31	56.71	48.32	48.21	48.16
9.42	56.79	56.97	48.51	48.41	48.36
9.50	57.81	57.25	48.72	48.62	48.57
9.58	58.32	57.36	48.91	48.82	48.77
9.67	58.83	57.47	49.12	49.02	48.98
9.75	59.38	57.53	49.28	49.18	49.14
9.83	58.22	57.64	49.47	49.38	49.34
9.92	58.47	57.75	49.59	49.5	49.46
10.00	60.4	58.26	49.72	49.64	49.6
10.08	60.54	58.18	50.36	50.28	50.24
10.17	60.69	58.39	50.72	50.65	50.61
10.25	60.88	58.61	51	50.92	50.89
10.33	60.47	58.76	51.2	51.13	51.1
10.42	60.88	58.93	51.36	51.29	51.25
10.50	61.3	59.39	51.47	51.4	51.37
10.56	61.36	59.26	51.53	51.46	51.43
10.61	61.42	59.43	51.59	51.52	51.49
10.68	61.5	59.63	51.67	51.6	51.56
10.75	61.58	59.84	51.74	51.67	51.64
10.83	61.81	60.18	51.79	51.72	51.69
10.92	62.08	60.55	51.82	51.76	51.72
11.00	57.34	60.83	51.83	51.76	51.73
11.08	57.45	60.98	51.87	51.8	51.77
11.17	57.51	61.08	51.84	51.77	51.74
11.25	57.57	61.2	51.79	51.73	51.69

Table A-10: Sacramento River - West Levee Freeboard: 200-year Downtown Area: I-5 to Deep Water Ship Channel

River Station	Freeboard (ft)		
(mi)	Existing	900-ft	1,800-ft
57.0026	9.22	10.23	10.78
57.2522	9.09	10.11	10.67
57.4972	8.91	9.93	10.48
57.6416			
57.8526			
58.0013			
58.2496			
58.4871	2.23	3.24	3.81
58.4892	2.19	3.21	3.78
58.5008	2.45	3.47	4.04
58.5029	2.45	3.47	4.04
58.5042	2.46	3.47	4.04
58.5063	2.46	3.47	4.04
58.5179	2.44	3.46	4.02
58.5199	2.44	3.46	4.02
58.7508	2.18	3.2	3.77
59.0028	0.13	1.15	1.72
59.2503	-0.06	0.96	1.53
59.2708	0.75	1.77	2.34
59.2729	0.68	1.7	2.28
59.2858	0.16	1.18	1.76
59.2878	-0.75	0.27	0.85
59.5005	-2.48	-1.46	-0.88
59.6782	2.86	3.9	4.47
59.6817	2.86	3.9	4.47
59.6838	2.86	3.89	4.47
59.6894			
59.6915	2.76	3.79	4.37
59.6963	2.61	3.64	4.22
59.7519	1.62	2.65	3.22
59.9979	25.22	26.26	26.84
60.2518	7.46	8.51	9.1
60.3926	7.18	8.22	8.81
60.3928	7.19	8.23	8.82

River Station	Freeboard (ft)		
(mi)	Existing	900-ft	1,800-ft
60.5957			
60.5958			
60.7522			
60.8811			
60.8813	-7.17	-6.12	-5.53
60.9982	5.73	6.78	7.37
60.9983	5.73	6.78	7.37
61.2547	5.01	6.07	6.65
61.4986	5.89	6.96	7.55
61.7455	5.21	6.28	6.87
62.0032	5.02	6.1	6.69
62.2564	5.03	6.11	6.71
62.4965	5.54	6.62	7.23
62.7519	5.69	6.78	7.39
62.9407	5.11	6.2	6.81
62.9428	5.11	6.2	6.81
62.9519	5.25	6.34	6.94
62.954	5.28	6.37	6.97
62.9596	5.34	6.43	7.03
62.9617	5.35	6.44	7.04
62.9708	5.36	6.45	7.06
62.9729	5.36	6.45	7.06
63.0067	5.33	6.42	7.03
63.2518	4.75	5.85	6.47
63.2696	4.73	5.84	6.45
63.2874	4.7	5.81	6.42
63.3052	4.69	5.79	6.41
63.323	4.66	5.76	6.38
63.3408	4.64	5.74	6.36
63.3586	4.61	5.71	6.33
63.3765	4.59	5.69	6.31
63.3943	4.57	5.68	6.3
63.4121	4.54	5.65	6.27
63.4299	4.52	5.63	6.25

River Station	Freeboard (ft)		
(mi)	Existing	900-ft	1,800-ft
63.4477	4.49	5.6	6.22
63.4655	4.47	5.58	6.2
63.4833	4.43	5.55	6.15
63.5012	4.39	5.5	6.11
63.5191	4.36	5.46	6.06
63.5371	4.33	5.41	6.02
63.5551	4.31	5.39	5.99
63.5731	4.3	5.37	5.97
63.5911	4.29	5.34	5.94
63.6091	4.29	5.34	5.92
63.6271	4.28	5.32	5.91
63.645	4.3	5.33	5.9
63.663	4.31	5.33	5.91
63.681	4.34	5.33	5.9
63.699	4.35	5.34	5.91
63.717	4.37	5.37	5.91
63.735	4.4	5.37	5.92
63.753	4.42	5.4	5.93
63.7718	4.45	5.45	5.96
63.7906	4.52	5.49	6.01
63.8095	4.55	5.54	6.04
63.8283	4.61	5.6	6.09
63.8471	4.64	5.62	6.13
63.866	4.63	5.68	6.16
63.8848	4.61	5.71	6.19
63.9037	4.6	5.74	6.23
63.9225	4.58	5.75	6.24
63.9413	4.57	5.76	6.26
63.9602	4.55	5.75	6.26
63.979	4.54	5.75	6.26
63.9979	4.52	5.74	6.25
64.2532	4.85	6.02	6.73
64.5033	4.89	6.04	6.73
64.7582	4.96	6.09	6.76

River Station	Freeboard (ft)		
(mi)	Existing	900-ft	1,800-ft
65.0046	4.89	5.99	6.64
65.2513	4.98	6.04	6.67
65.5045	4.98	5.99	6.59
65.7539	5.64	6.64	7.23
66.0073	5.38	6.35	6.92
66.2529	4.39	5.35	5.91
66.5012	3.85	4.75	5.27
66.7511	4.1	4.97	5.48
67.0013	4.11	4.92	5.4
67.2532	3.41	4.22	4.69
67.5018	3.29	4.08	4.53
67.7513	3.01	3.77	4.2
68.0022	2.82	3.53	3.94
68.2473	2.58	3.26	3.65
68.5059	1.47	2.11	2.48
68.7477	1.76	2.39	2.75
69.0004	1.99	2.59	2.94
69.2511	1.43	2.01	2.35
69.5003	1.84	2.41	2.74
69.7512	1.79	2.34	2.66
69.9983	1.67	2.2	2.51
70.2464	1.79	2.31	2.61
70.4988	2.1	2.61	2.9
70.7462	2.38	2.88	3.16
70.8909	1.74	2.22	2.5
70.9239	1.77	2.25	2.52
71.0034	1.7	2.17	2.43
71.2551	2.38	2.82	3.07
71.5027	1.81	2.23	2.46
71.7536	1.73	2.13	2.36
71.9985	0.9	1.29	1.51
72.253	0.6	0.95	1.16
72.5009	1.89	2.22	2.42

Table A-11: Sacramento River - East Levee Freeboard: 200-year Downtown Area: I-5 to Deep Water Ship Channel

Rvr Sta	Levee Elevation (ft)		WSEL (ft)		
(mi)	West	East	Existing	900-Ft	1,800-Ft
40.0021	30.28	30.67	26.39	25.71	25.35
40.2518	31.26	29.89	26.62	25.93	25.56
40.5017	31.94	30.58	26.81	26.11	25.73
40.7534	31.42	28.94	26.97	26.26	25.88
41.0011	32.13	28.21	27.17	26.45	26.06
41.2483	30.78	29.49	27.3	26.59	26.19
41.4991	30.11	29.8	27.48	26.75	26.35
41.7493	31.72	30.07	27.58	26.85	26.45
41.9982	34.19	31.04	27.77	27.03	26.62
42.25	32.34	32.67	27.95	27.2	26.79
42.5014	31.8	30.77	28.04	27.29	26.87
42.7493	31.74	32.22	28.11	27.36	26.95
43.0002	33.05	32.98	28.13	27.38	26.97
43.2488	31.98	33.04	28.31	27.55	27.14
43.5022	32.08	32.69	28.46	27.7	27.28
43.7416	32.78	33.27	28.78	27.99	27.56
43.9965	32.56	32.88	28.83	28.04	27.61
44.2514	32.3	33.1	28.85	28.07	27.64
44.4976	32.12	32.79	28.97	28.19	27.75
44.7498	33.33	34.46	29.18	28.38	27.94
45.001	34.21	34.79	29.3	28.49	28.05
45.2497	34.54	34.46	29.48	28.67	28.22
45.5037	35.46	35.02	29.69	28.87	28.41
45.7528	31.6	34.71	29.83	29	28.54
46.0043	32.28	35.37	29.93	29.1	28.64
46.2507	32.29	34.73	30.05	29.21	28.75
46.4177	34.58	34.8	30.18	29.33	28.86
46.4198	34.82	34.53	30.17	29.33	28.86
46.4247	34.98	34.51	30.2	29.36	28.89
46.4268	34.92	34.48	30.2	29.36	28.89
46.502	33.09	33.52	30.24	29.39	28.92
46.7517	32.87	34.33	30.25	29.4	28.93
47.0021	33.75	34.13	30.28	29.43	28.97

Rvr Sta	Levee Elevation (ft)		WSEL (ft)		
(mi)	West	East	Existing	900-Ft	1,800-Ft
47.2501	32.64	34.58	30.42	29.56	29.09
47.504	35.72	35.52	30.71	29.83	29.35
47.7538	33.93	35.27	30.84	29.96	29.48
48.0025	33.11	34.42	30.97	30.09	29.61
48.2539	34.38	34.69	31.03	30.15	29.66
48.502	34.85	32.92	31.12	30.23	29.75
48.7539	34.06	33.71	31.21	30.33	29.84
49	34.69	34.06	31.27	30.39	29.9
49.2511	34.99	34.16	31.45	30.56	30.07
49.5012	35.69	35.02	31.62	30.72	30.22
49.7528	35.66	36.71	31.74	30.84	30.34
49.9997	34.66	36.27	31.95	31.03	30.53
50.2486	34.55	33.87	32.02	31.1	30.59
50.4987	35.05	33.8	32.1	31.18	30.68
50.7537	35.29	35.2	32.15	31.23	30.72
51.0016	35.97	36.48	32.14	31.23	30.72
51.2471	36.04	38.73	32.28	31.36	30.85
51.5034	34.92	38.41	32.5	31.57	31.06
51.7539	34.59	39.64	32.62	31.69	31.17
52.0027	36.44	41.27	32.63	31.7	31.19
52.2547	36.55	40.53	32.73	31.8	31.28
52.4994	36.18	38.62	33.05	32.1	31.58
52.7474	35.86	40.03	33.2	32.24	31.71
53.0006	36.82	40.66	33.26	32.31	31.78
53.2494	36.44	40.76	33.31	32.35	31.83
53.5013	36.75	39.9	33.3	32.35	31.83
53.7528	34.68	39.4	33.46	32.5	31.98
54.0001	37.06	41.87	33.59	32.63	32.1
54.2524	37.08	41.05	33.8	32.82	32.38
54.499	40.06	41.26	33.95	32.97	32.43
54.7464	40.58	39.21	34.11	33.12	32.58
55.0013	39.14	39.33	34.28	33.29	32.74
55.248	39.28	40.6	34.25	33.26	32.72

Rvr Sta	Levee Elevation (ft)		WSEL (ft)		
(mi)	West	East	Existing	900-Ft	1,800-Ft
55.4949	38.82	39.47	34.33	33.34	32.79
55.7498	49.39	38.5	34.52	33.53	32.97
56.0029	40.14	41.82	34.47	33.48	32.94
56.2531	45.08	40.43	34.6	33.61	33.06
56.5018	44.28	41.25	34.77	33.77	33.22
56.7515	44.4	41.74	34.91	33.9	33.35
57.0026	44.23	41.74	35.01	34	33.45
57.2522	44.2	40.89	35.11	34.09	33.53
57.4972	44.07	41.38	35.16	34.14	33.59
57.6416		42.54	35.29	34.26	33.7
57.8526			35.44	34.41	33.84
58.0013			35.5	34.43	33.87
58.2496			35.59	34.57	34.01
58.4871	37.9	39.54	35.67	34.66	34.09
58.4892	37.87	38.43	35.68	34.66	34.09
58.5008	38.13	39.64	35.68	34.66	34.09
58.5029	38.13	39.65	35.68	34.66	34.09
58.5042	38.13	39.65	35.67	34.66	34.09
58.5063	38.13	39.66	35.67	34.66	34.09
58.5179	38.13	39.7	35.69	34.67	34.11
58.5199	38.13	30.71	35.69	34.67	34.11
58.7508	37.99	40.46	35.81	34.79	34.22
59.0028	36.06	40.34	35.93	34.91	34.34
59.2503	36	40.07	36.06	35.04	34.47
59.2708	36.81	40.29	36.06	35.04	34.47
59.2729	36.74	40.68	36.06	35.04	34.46
59.2858	36.26		36.1	35.08	34.5
59.2878	35.35		36.1	35.08	34.5
59.5005	33.58	40.66	36.06	35.04	34.46
59.6782	39.09		36.23	35.19	34.62
59.6817	39.09		36.23	35.19	34.62
59.6838	39.09		36.23	35.2	34.62
59.6894			36.24	35.21	34.63

Rvr Sta	Levee Elevation (ft)		WSEL (ft)		
(mi)	West	East	Existing	900-Ft	1,800-Ft
59.6915	39	38.41	36.24	35.21	34.63
59.6963	38.85	38.98	36.24	35.21	34.63
59.7519	37.82	40.41	36.2	35.17	34.6
59.9979	61.62	42.4	36.4	35.36	34.78
60.2518	44.02	42.93	36.56	35.51	34.92
60.3926	43.74	42.89	36.58	35.52	34.93
60.3928	43.75	42.87	36.56	35.52	34.93
60.5957		35.67	36.56	35.52	34.93
60.5958		35.68	36.58	35.52	34.93
60.7522		45.52	36.57	35.52	34.93
60.8811		45.79	36.57	35.52	34.93
60.8813	29.4	45.79	36.57	35.52	34.93
60.9982	42.3	44.02	36.57	35.52	34.93
60.9983	42.3	44.02	36.57	35.52	34.93
61.2547	41.56	43.39	36.55	35.49	34.91
61.4988	42.44	43.2	36.55	35.48	34.89
61.7455	41.75	43.07	36.54	35.47	34.88
62.0032	41.56	44.03	36.54	35.46	34.87
62.2564	41.56	41.32	36.53	35.45	34.85
62.4965	42.06	39.43	36.52	35.44	34.83
62.7519	42.2	43.16	36.51	35.42	34.81
62.9407	41.63	41.24	36.52	35.43	34.82
62.9428	41.63	41.28	36.52	35.43	34.82
62.9519	41.76	41.41	36.51	35.42	34.82
62.954	41.79	41.44	36.51	35.42	34.82
62.9596	41.85	41.46	36.51	35.42	34.82
62.9617	41.86	41.43	36.51	35.42	34.82
62.9708	41.87	41.33	36.51	35.42	34.81
62.9729	41.87	41.31	36.51	35.42	34.81
63.0067	41.84	41.18	36.51	35.42	34.81
63.2518	41.25	40.43	36.5	35.4	34.78
63.2696	41.23	40.43	36.5	35.39	34.78
63.2874	41.2	40.43	36.5	35.39	34.78

Rvr Sta	Levee Elevation (ft)		WSEL (ft)		
(mi)	West	East	Existing	900-Ft	1,800-Ft
63.3052	41.18	40.43	36.49	35.39	34.77
63.323	41.15	40.43	36.49	35.39	34.77
63.3408	41.13	40.43	36.49	35.39	34.77
63.3586	41.1	40.43	36.49	35.39	34.77
63.3765	41.08	40.43	36.49	35.39	34.77
63.3943	41.06	40.43	36.49	35.38	34.76
63.4121	41.03	40.43	36.49	35.38	34.76
63.4299	41.01	40.43	36.49	35.38	34.76
63.4477	40.98	40.43	36.49	35.38	34.76
63.4655	40.96	40.43	36.49	35.38	34.76
63.4833	40.93	40.43	36.5	35.38	34.78
63.5012	40.91	40.43	36.52	35.41	34.8
63.5191	40.89	40.44	36.53	35.43	34.83
63.5371	40.86	40.45	36.53	35.45	34.84
63.5551	40.84	40.46	36.53	35.45	34.85
63.5731	40.82	40.47	36.52	35.45	34.85
63.5911	40.79	40.48	36.5	35.45	34.85
63.6091	40.77	40.49	36.48	35.43	34.85
63.6271	40.74	40.51	36.46	35.42	34.83
63.645	40.72	40.52	36.42	35.39	34.82
63.663	40.7	40.53	36.39	35.37	34.79
63.681	40.67	40.54	36.33	35.34	34.77
63.699	40.65	40.55	36.3	35.31	34.74
63.717	40.63	40.56	36.26	35.26	34.72
63.735	40.6	40.57	36.2	35.23	34.68
63.753	40.58	40.58	36.16	35.18	34.65
63.7718	40.57	40.41	36.12	35.12	34.61
63.7906	40.57	40.25	36.05	35.08	34.56
63.8095	40.56	40.08	36.01	35.02	34.52
63.8283	40.56	39.92	35.95	34.96	34.47
63.8471	40.55	39.75	35.91	34.93	34.42
63.866	40.55	39.59	35.92	34.87	34.39
63.8848	40.54	39.42	35.93	34.83	34.35

Rvr Sta	Levee Elevation (ft)		WSEL (ft)		
(mi)	West	East	Existing	900-Ft	1,800-Ft
63.9037	40.54	39.26	35.94	34.8	34.31
63.9225	40.53	39.09	35.95	34.78	34.29
63.9413	40.53	38.93	35.96	34.77	34.27
63.9602	40.52	38.76	35.97	34.77	34.26
63.979	40.52	38.6	35.98	34.77	34.26
63.9979	40.51	38.43	35.99	34.77	34.26
64.2532	41.04	39.16	36.19	35.02	34.31
64.5033	41.15	39.43	36.26	35.11	34.42
64.7562	41.33	40.82	36.37	35.24	34.57
65.0046	41.41	40.39	36.52	35.42	34.77
65.2513	41.68	40.81	36.7	35.64	35.01
65.5045	41.86	41.49	36.88	35.87	35.27
65.7539	42.58	41.63	36.94	35.94	35.35
66.0073	42.39	42.02	37.01	36.04	35.47
66.2529	41.53	42.55	37.14	36.18	35.62
66.5012	41.26	43.61	37.41	36.51	35.99
66.7511	41.67	40.62	37.57	36.7	36.19
67.0013	41.95	42.01	37.84	37.03	36.55
67.2532	41.32	41.15	37.91	37.1	36.63
67.5018	41.32	39.95	38.03	37.24	36.79
67.7513	41.2	41.02	38.19	37.43	37
68.0022	41.25	40.42	38.43	37.72	37.31
68.2473	41.19	40.39	38.61	37.93	37.54
68.5059	40.32	41.39	38.85	38.21	37.84
68.7477	40.68	41.48	38.92	38.29	37.93
69.0004	41.1	41.4	39.11	38.51	38.16
69.2511	40.65	42.62	39.22	38.64	38.3
69.5003	41.13	42.52	39.29	38.72	38.39
69.7512	41.16	42.11	39.37	38.82	38.5
69.9983	41.21	42.27	39.54	39.01	38.7
70.2464	41.38	42	39.59	39.07	38.77
70.4988	41.76	42.92	39.66	39.15	38.86
70.7462	42.13	43.06	39.75	39.25	38.97

Rvr Sta	Levee Elevation (ft)		WSEL (ft)		
(mi)	West	East	Existing	900-Ft	1,800-Ft
70.8909	41.58	42.83	39.82	39.34	39.06
70.9239	41.62	42.82	39.85	39.37	39.1
71.0034	41.64	42.77	39.94	39.47	39.21
71.2551	42.53	43.19	40.15	39.71	39.46
71.5027	42.1	41.88	40.29	39.87	39.64
71.7536	42.12	42.94	40.39	39.99	39.76
71.9985	41.33	42.64	40.43	40.04	39.82
72.253	41.32	42.61	40.72	40.37	40.16
72.5009	42.76	44.45	40.87	40.54	40.34
72.7504	42.23	43.78	41.07	40.77	40.59
73.0002	43.08	43.99	41.17	40.88	40.71
73.2532	43.02	41.53	41.35	41.07	40.92
73.4993	42.36	43.4	41.43	41.18	41.03
73.7495	43.16	43.27	41.52	41.29	41.16
73.9998	42.36	41.89	41.62	41.4	41.28
74.2531	43.19	41.1	41.74	41.53	41.41
74.5028	43.49	41.21	41.79	41.59	41.47
74.7528	44.19	43.03	42.08	41.9	41.8
75.0012	43.42	41.76	42.21	42.05	41.95
75.2546	42.36	41.57	42.3	42.15	42.05
75.5022	42.4	42.15	42.46	42.31	42.23
75.7537	42.66	41.38	42.6	42.47	42.39
76.0059	43.08	43.35	42.71	42.58	42.51
76.2542	42.89	42.09	42.88	42.77	42.71
76.5016	43.26	41.22	43.02	42.92	42.86
76.7497	43.58	43.01	43.1	43	42.94
77.0052	43.42	42.69	43.24	43.15	43.09
77.2512	43.97	42.46	43.4	43.31	43.26
77.5013	44.06	43.84	43.45	43.36	43.31
77.7511	44.41	45.72	43.71	43.64	43.59
78.0019	44.93	45.27	43.93	43.87	43.84
78.2548	44.87	43.89	44.01	43.95	43.92
78.5025	45.05	45.42	44.14	44.09	44.06

Rvr Sta	Levee Elevation (ft)		WSEL (ft)		
(mi)	West	East	Existing	900-Ft	1,800-Ft
78.7536	44.06	43.48	44.31	44.27	44.24
79.0022	44.21	48.44	44.46	44.42	44.4
79.203	46.22	48	44.62	44.59	44.57
79.2049	46.24	48	44.63	44.59	44.57
79.32	46.16	45.4	44.63	44.59	44.57
79.33	46.16	45.4	44.63	44.59	44.57
79.5	47.01	46.26	44.8	44.77	44.75
79.75	47.96	47.37	44.92	44.89	44.88
79.999	46.35	45.6	44.97	44.94	44.93
80	46.28	45.6	44.97	44.94	44.93
80.39		46.63	44.97	44.94	44.93
80.49		46.82	45.04	45.02	45.01
80.75		47.78	45.07	45.05	45.04
81.26		45.56	45.07	45.06	45.04
81.5		44.96	45.07	45.06	45.05
81.75		46.19	45.08	45.06	45.05
81.99		47.55	44.92	44.9	44.89
82.26			44.91	44.89	44.88
82.5			44.95	44.93	44.92
82.74			44.94	44.93	44.92
83			44.97	44.95	44.94
83.25			44.94	44.92	44.91
83.5			44.98	44.96	44.95
83.75	47.35		44.61	44.6	44.59
84	48.4		44.64	44.63	44.62
84.25	47.81		44.63	44.62	44.61
84.5	47.17		44.69	44.68	44.67
84.75	47.21	47.94	44.69	44.68	44.67
85	47.1	48.12	44.69	44.68	44.67
85.25	45.47	47.62	44.69	44.68	44.67
85.5	45.06	47.81	44.69	44.68	44.67
85.75	45.08	47.21	44.7	44.68	44.68
86	45.16	45.7	44.7	44.68	44.68

Rvr Sta	Levee Elevation (ft)		WSEL (ft)		
(mi)	West	East	Existing	900-Ft	1,800-Ft
86.25	44.4	43.7	44.7	44.68	44.68
86.5	44.71	44.94	44.67	44.66	44.65
86.75	44.57	44.28	44.65	44.64	44.63
87	45.16	45.16	44.63	44.62	44.62
87.25	45.55	44.73	44.65	44.64	44.63
87.5	45.23	47.31	44.69	44.68	44.68
87.75	46.23	46.41	44.66	44.65	44.65
88	46.7	50	44.79	44.78	44.78
88.25	46.37	45.65	44.8	44.79	44.79
88.5	46.38	44.46	44.79	44.78	44.78
88.75	46.69	45.12	44.79	44.78	44.77
89	46.98	50.91	44.84	44.84	44.83
89.25	46.08	46.6	44.86	44.85	44.85
89.5	48.13	45.51	44.91	44.91	44.9
89.75	47.6	46.73	44.94	44.93	44.93
90	47.47	46.13	45	44.99	44.99
90.22	46.24	46.41	44.98	44.97	44.97
90.222	46.26	46.3	44.97	44.97	44.96
90.229	46.7	45.11	45.04	45.04	45.03
90.231	46.81	44.98	45.06	45.05	45.05
90.25	47.4	45.48	45.09	45.08	45.08
90.3	47.6	45.69	45.12	45.12	45.11
90.36	47.35	47.37	45.13	45.12	45.12
90.5	47.58	47	45.16	45.15	45.15
90.75	47.6	46.57	45.19	45.18	45.18
91	47.79	47	45.26	45.25	45.25

Table A-12: 200-year Water Surface Profile: Yolo Bypass

Rvr Sta	Levee Elevation (ft)		WSEL (ft)		
(mi)	West	East	Existing	900-Ft	1,800-Ft
15.208	20.84	10.54	11.86	11.85	11.85
15.2081	20.84	10.54	11.86	11.85	11.85
15.457	20.99	11.15	11.99	11.98	11.98
15.708	21.92	12.87	13.48	13.49	13.5
15.957	21.78	11.39	15.34	15.35	15.38
16.207	24.43	11.3	16.06	16.07	16.1
16.461	21.18	11.42	16.61	16.63	16.66
16.707	22.34	11.6	16.86	16.88	16.91
16.956	20.87	11.87	17.03	17.05	17.08
17.208	20.72	11.75	17.22	17.24	17.27
17.457	21.82	11.89	17.59	17.62	17.64
17.708	21.9	11.65	18.09	18.12	18.15
17.958	22.74	13.14	18.42	18.45	18.48
18.208	22.62	14.22	19.2	19.22	19.25
18.4738	20.96	14.97	20.49	20.52	20.55
18.474	20.96	14.97	20.49	20.52	20.55
18.767	11.71	24.53	20.49	20.52	20.55
18.7671	11.71	24.53	20.49	20.52	20.55
18.974	8.44	23.38	20.54	20.56	20.59
19.161	11.32	22.66	20.59	20.62	20.65
19.3478	10.49	24.46	20.66	20.69	20.72
19.348	10.49	24.46	20.66	20.69	20.72
19.534	10.07	20.55	20.66	20.69	20.72
19.5341	10.07	20.55	20.66	20.69	20.72
19.721	9.39	21.63	20.7	20.73	20.76
19.907	9.97	19.76	20.75	20.78	20.81
20.093	15.02	20.93	20.79	20.82	20.85
20.28	17.99	21.5	20.82	20.86	20.88
20.466	19.02	21.47	20.86	20.89	20.92
20.6528	18.73	19.58	20.91	20.94	20.97
20.653	18.73	19.58	20.91	20.94	20.97
20.839	18.2	20.03	20.91	20.94	20.97
20.8391	18.2	20	20.91	20.94	20.97

Rvr Sta	Levee Elevation (ft)		WSEL (ft)		
(mi)	West	East	Existing	900-Ft	1,800-Ft
21.025	16.74	22.79	20.93	20.97	21
21.212	19.34	21.91	21.01	21.05	21.08
21.398	19.11	23.55	21.07	21.1	21.13
21.585	19.46	18.82	21.13	21.17	21.2
21.771	18.98	22.04	21.21	21.24	21.28
21.958	20.02	22.59	21.27	21.31	21.34
22.144	20.64	23.11	21.34	21.38	21.41
22.331	20.22	23.17	21.43	21.47	21.5
22.518	28.72	23.96	21.52	21.56	21.6
22.705	28.21	23.56	21.63	21.67	21.7
22.892	27.5	24.34	21.73	21.77	21.81
23.079	27.7	23.7	21.83	21.87	21.91
23.266	28.79	24.21	21.92	21.96	22
23.453	28.59	24.57	22.02	22.07	22.11
23.64	26.72	25.01	22.11	22.16	22.2
23.826	26.69	25.87	22.19	22.24	22.28
24.013	27.73	25.88	22.3	22.35	22.4
24.2	27.52	25.1	22.41	22.46	22.5
24.387	26.94	25.47	22.51	22.57	22.61
24.574	28.1	26.22	22.61	22.66	22.71
24.761	28.07	25.65	22.72	22.77	22.82
24.948	28.05	24.71	22.83	22.88	22.93
25.135	28.74	25	22.94	23	23.05
25.322	27.79	25.35	23.05	23.11	23.16
25.508	27.9	26.47	23.15	23.22	23.27
25.695	27.65	27.45	23.26	23.34	23.39
25.883	27.8	27.29	23.4	23.48	23.53
26.071	27.82	27.32	23.51	23.59	23.64
26.259	27.74	27.27	23.63	23.71	23.76
26.447	25.65	27.61	23.75	23.84	23.89
26.635	27.95	27.55	23.86	23.95	24
26.823	27.9	27.15	23.95	24.04	24.09
27.011	28.58	27.28	24.04	24.13	24.18

Rvr Sta	Levee Elevation (ft)		WSEL (ft)		
(mi)	West	East	Existing	900-Ft	1,800-Ft
27.199	27.83	27.48	24.12	24.21	24.27
27.387	29.21	28.38	24.2	24.29	24.35
27.575	29.12	28.47	24.28	24.37	24.43
27.763	29.02	28.41	24.38	24.47	24.53
27.951	30.31	28.78	24.47	24.57	24.63
28.139	29.71	28.53	24.58	24.67	24.73
28.327	28.73	28.41	24.7	24.79	24.85
28.515	28.86	26.76	24.84	24.94	25
28.703	29.01	27.36	24.99	25.08	25.14
28.891	29.68	28.32	25.15	25.25	25.3
29.079	29.43	27.08	25.45	25.55	25.61
29.267	31.3	28.28	25.72	25.81	25.87
29.455	29.2	25.67	25.97	26.06	26.12
29.643	28.61	28.56	26.16	26.25	26.31
29.835	28.89	28.81	26.29	26.39	26.45
30.026	29.3	28.96	26.42	26.51	26.57
30.217	29.08	29.71	26.55	26.64	26.7
30.408	28.16	25.73	26.67	26.77	26.83
30.599	28.96	26.89	26.79	26.89	26.95
30.79	29.08	24.83	26.91	27	27.06
30.981	29.14	29.27	27.01	27.1	27.17
31.172	31.89	29.71	27.13	27.23	27.29
31.363	32.05	29.74	27.35	27.44	27.5
31.554	32.52	30.89	27.56	27.65	27.71
31.745	33.03	32.01	27.72	27.82	27.88
31.936	32.83	33.28	27.86	27.96	28.02
32.131	33.13	33.43	27.99	28.09	28.15
32.325	33.7	33.7	28.13	28.23	28.29
32.52	32.94	32.73	28.28	28.38	28.44
32.714	30.44	31.02	28.42	28.52	28.58
32.909	31.09	30.18	28.52	28.62	28.68
33.103	31	30.52	28.61	28.71	28.77
33.298	31.81	26.21	28.7	28.8	28.86

Rvr Sta (mi)	Levee Elevation (ft)		WSEL (ft)		
	West	East	Existing	900-Ft	1,800-Ft
33.492	32	31.41	28.79	28.89	28.95
33.687	32.16	32.67	28.87	28.97	29.04
33.882	31.76	34.21	28.95	29.06	29.11
34.079	31.72	32.47	29.02	29.12	29.18
34.276	32.01	33.77	29.08	29.19	29.25
34.474	31.58	32.88	29.15	29.25	29.31
34.671	31.7	34.31	29.21	29.31	29.38
34.869	31.09	32.82	29.28	29.38	29.45
35.066	31.22	32.14	29.35	29.46	29.52
35.268	31.52	34.18	29.44	29.54	29.6
35.47	30.83	31.61	29.52	29.62	29.69
35.672	29.75	32.11	29.61	29.71	29.78
35.874	31.79	33.73	29.71	29.81	29.87
36.077	32.16	33.03	29.81	29.92	29.98
36.279	32.06	33.95	29.91	30.01	30.08
36.487	32.47	32.47	29.99	30.09	30.16
36.695	32.09	33.4	30.07	30.17	30.23
36.904	31.75	33.91	30.14	30.24	30.31
37.112	32.95	33.54	30.21	30.32	30.38
37.32	33.14	37.52	30.28	30.39	30.45
37.529	32.36	38.94	30.35	30.46	30.52
37.727	33.22	34.59	30.43	30.54	30.6
37.926	32.58	38.72	30.54	30.64	30.71
38.125	33.05	33.24	30.65	30.75	30.82
38.324	33.42	38.28	30.76	30.86	30.93
38.5218	33.53		30.91	31.02	31.08
38.522	33.53		30.93	31.03	31.1
38.656	33.02		30.93	31.03	31.1
38.6561	33.02		30.93	31.03	31.1
38.813	32.65		31.03	31.13	31.2
38.938	33.68		31.11	31.22	31.29
39.084	33.51		31.16	31.27	31.34
39.271	33.78	33.22	31.29	31.4	31.46

Rvr Sta (mi)	Levee Elevation (ft)		WSEL (ft)		
	West	East	Existing	900-Ft	1,800-Ft
39.458	33.62	32.58	31.39	31.5	31.57
39.645	33.72	32.3	31.48	31.59	31.66
39.831	33.69	34.09	31.57	31.68	31.74
40.018	33.1	33.24	31.65	31.76	31.83
40.204	33.75	33.41	31.71	31.82	31.89
40.39	33.89	33.37	31.78	31.89	31.96
40.577	34.44	32.32	31.85	31.96	32.03
40.763	35.57	33.04	31.91	32.02	32.09
40.95	35.64	32.83	31.97	32.09	32.16
41.136	34.76	33.78	32.04	32.15	32.22
41.322	33.27	33.68	32.09	32.2	32.27
41.509	33.61	33.65	32.15	32.26	32.33
41.695	33.34	33.51	32.2	32.32	32.39
41.882	33.45	33.71	32.25	32.37	32.44
42.11	34.04	33.72	32.31	32.43	32.5
42.342	33.29	33.53	32.36	32.48	32.55
42.51	33.47	35.71	32.61	32.73	32.81
42.62	33.6	34.9	32.64	32.77	32.84
42.6883	33.6	34.61	32.65	32.78	32.85
42.7586	33.6	34.33	32.86	32.99	33.07
42.828	33.6	34.04	32.87	33	33.08
42.989	33.6	36.81	32.87	33	33.08
42.9891	33.6	36.81	32.87	33	33.08
43.089	33.6	36.84	32.89	33.02	33.1
43.375	33.55	35.6	32.94	33.07	33.15
43.582	33.55	37.15	32.98	33.11	33.19
43.751	33.65	36.54	33.01	33.15	33.23
43.9488	33.88	36.5	33.06	33.19	33.27
43.949	33.88	36.5	33.06	33.19	33.27
44.13	34.98	34.98	33.06	33.19	33.27
44.1301	34.98	34.98	33.06	33.19	33.27
44.498	32.32	35.74	33.11	33.25	33.33
44.81	33.31	34.52	33.18	33.31	33.38

Rvr Sta (mi)	Levee Elevation (ft)		WSEL (ft)		
	West	East	Existing	900-Ft	1,800-Ft
45.059	33.15	34.82	33.23	33.36	33.43
45.37	33.44	34.9	33.31	33.44	33.51
45.598	33.81	35.96	33.38	33.5	33.57
45.829	34.36	35.96	33.45	33.57	33.64
46.245	33.44	35.91	33.59	33.71	33.77
46.452	33.32	37.22	33.67	33.78	33.85
46.658	34.3	36.21	33.75	33.86	33.93
46.857	35.14	36.42	33.83	33.94	34
47.076	34.94	34.88	33.93	34.03	34.09
47.279	35.68	36.62	34.02	34.12	34.18
47.488	36.78	37.44	34.1	34.2	34.26
47.711	35.88	37.44	34.2	34.3	34.35
47.92	35.61	36.9	34.29	34.38	34.44
48.124	36.29	35.86	34.36	34.46	34.51
48.327	35.47	33.82	34.44	34.54	34.59
48.558	36.06	35.43	34.52	34.61	34.67
48.84	37.7	35.66	34.63	34.72	34.77
49.105	36.89	38.08	34.77	34.85	34.9
49.385	37.94	37.68	34.96	35.04	35.09
49.653	37.78	38.4	35.17	35.24	35.29
50.084	38.63	38.97	35.52	35.59	35.63
50.2895	36.69	38.85	35.76	35.83	35.86
50.495	34.75	38.74	35.94	36	36.03
50.496	34.75	38.74	35.94	36	36.03
50.503	36.82	42.38	36.01	36.06	36.09
50.63	38.9	46.01	36.37	36.42	36.44
50.697	40.97	49.65	36.43	36.48	36.5
50.911	38.66	49.98	36.61	36.65	36.68
51.097	39.99	34.89	36.79	36.83	36.85
51.284	39.85	50.55	36.98	37.02	37.04
51.47	40.02	48.69	37.12	37.16	37.18
51.657	41.58	49.4	37.28	37.32	37.34
51.844	40.68	49.4	37.44	37.47	37.49

Rvr Sta (mi)	Levee Elevation (ft)		WSEL (ft)		
	West	East	Existing	900-Ft	1,800-Ft
52.043	40.98	50.68	37.59	37.62	37.64
52.243	42.02	42.4	37.71	37.73	37.75
52.442	41.13	42.14	37.81	37.84	37.85
52.642	40.67	41.25	37.91	37.94	37.95
53.041	40.34	41.78	38.12	38.15	38.16
53.24	41.31	42.4	38.22	38.25	38.26
53.44	42.88	42.04	38.32	38.34	38.36
53.533	42.64	41.17	38.37	38.39	38.4
53.6258	42.45	40.36	38.41	38.43	38.45
53.626	42.4	40.31	38.41	38.44	38.45
53.813	42.63	40.9	38.41	38.44	38.45
53.8132	42.63	40.9	38.44	38.47	38.48
53.999	43.02	42.44	38.51	38.53	38.54
54.186	42.3	42.38	38.61	38.62	38.63
54.372	42.63	43.59	38.73	38.75	38.76
54.558	44.36	44.1	38.85	38.87	38.88
54.745	43.51	44.5	38.98	39	39
54.931	43.44	44.33	39.12	39.14	39.14
55.118	43.84	44.2	39.26	39.28	39.28
55.304	43.36	44.9	39.46	39.47	39.47
55.49	44.54	44.81	39.68	39.69	39.7
55.677	44.74	43.2	39.94	39.95	39.95
55.863	44.21	44.9	40.24	40.24	40.24
56.05	45.06	45.58	40.49	40.49	40.48
56.236	44.8	45.07	40.8	40.81	40.81
56.629	44.85	45.07	41.55	41.55	41.55

Table A-13: 200-year Water Surface Profile: Sacramento Bypass

Rvr Sta	Levee Elevation (ft)		WSEL (ft)		
(mi)	South	North	Existing	900-Ft	1,800-Ft
0		36.03	33.06	33.19	33.27
0.000183		36.03	33.06	33.19	33.27
0.18	35.06	34.28	33.11	33.23	33.3
0.37	34.82	35.17	33.37	33.35	33.38
0.56	35.39	36.27	33.62	33.49	33.45
0.75	37.91	36.63	33.85	33.6	33.52
0.93	36.06	37.14	34.13	33.73	33.59
1.12	37.36	36.35	34.42	33.88	33.68
1.3	36.87	38.04	34.65	34.02	33.77
1.49	36.82	38.4	34.98	34.23	33.9
1.67981	39.66	40.9	35.42	34.48	34.06
1.68	39.66	40.9	36.25	35.02	34.43

Table A-14: 200-year Water Surface Profile: American River

Rvr Sta	Levee Elevation (ft)		WSEL (ft)		
(mi)	South	North	Existing	900-Ft	1,800-Ft
0.115	43.37		36.56	35.52	34.93
0.182	45.16		36.71	35.68	35.11
0.184	45.52		36.72	35.69	35.12
0.192	44.71		36.82	35.8	35.23
0.194	44.39		36.82	35.8	35.24
0.226	53.56		36.8	35.78	35.22
0.256	55.35		36.95	35.95	35.41
0.258	52.9		36.95	35.96	35.41
0.271	53.09		36.99	36	35.46
0.273	56.06		36.98	36	35.46
0.274	55.65		36.97	35.99	35.44
0.289	55.84		37.02	36.04	35.5
0.29	56.18		37.01	36.03	35.49
0.36	47.09		37.06	36.09	35.55
0.43	41.48		37.18	36.22	35.7
0.5	41.42		37.27	36.33	35.82
0.583333	41.45		37.36	36.44	35.93
0.666666	41.48		37.44	36.53	36.03
0.75	41.57		37.52	36.61	36.12
0.833333	41.48		37.62	36.72	36.24
0.916666	41.29		37.7	36.81	36.34
1	41.1		37.79	36.92	36.45
1.08333	41.05		37.89	37.03	36.56
1.16666	41.2		37.99	37.14	36.69
1.25	41.35		38.08	37.24	36.8
1.33333	41.42		38.11	37.28	36.83
1.41666	41.35		38.17	37.35	36.91
1.5	42.09		38.23	37.41	36.98
1.58133	41.47		38.27	37.46	37.02
1.66266	41.65		38.27	37.45	37.02
1.744	41.83		38.28	37.46	37.03
1.745	41.83		38.27	37.45	37.02
1.753	41.85		38.28	37.47	37.03

Rvr Sta	Levee Elevation (ft)		WSEL (ft)		
(mi)	South	North	Existing	900-Ft	1,800-Ft
1.754	41.85		38.28	37.46	37.03
1.81933	42.07		38.36	37.55	37.13
1.88466	43.3	43.06	38.45	37.65	37.23
1.95	44.55	43.8	38.51	37.73	37.32
1.98	47.38	43.47	38.59	37.82	37.42
1.99	45.23	43.49	38.72	37.97	37.57
2.01	49.1	43.53	38.81	38.06	37.67
2.02	49.1	43.55	38.82	38.07	37.68
2.04	46	43.78	38.85	38.11	37.72
2.0681	45.85	43.66	38.9	38.16	37.77
2.07	45.75	43.66	38.9	38.16	37.78
2.0722	45.72	43.67	38.99	38.25	37.86
2.0741	45.63	43.67	39	38.26	37.87
2.164	45.65	43.87	39.16	38.43	38.05
2.165	45.7	43.87	39.16	38.44	38.06
2.168	45.65	43.87	39.22	38.5	38.12
2.169	45.7	43.88	39.22	38.5	38.12
2.25	44.7	44.05	39.29	38.58	38.2
2.266	45	44.09	39.34	38.63	38.26
2.267	44.7	44.1	39.27	38.55	38.17
2.271	44.7	44.1	39.31	38.6	38.22
2.272	44.7	44.1	39.3	38.58	38.21
2.348	44.71	44.27	39.57	38.87	38.5
2.424	44.99	44.43	39.65	38.96	38.6
2.5	45	44.57	39.72	39.04	38.68
2.58333	45.01	44.73	39.81	39.13	38.78
2.66666	45.02	44.9	39.88	39.21	38.86
2.75	46.48	45.05	39.95	39.29	38.94
2.83333	57.25	45.2	39.98	39.32	38.97
2.91666	67.9	45.32	40.01	39.36	39.02
3	78.56	45.42	40.06	39.41	39.07
3.08333	80.01	45.51	40.1	39.45	39.11
3.16666	81.46	45.64	40.14	39.5	39.16

Rvr Sta	Levee Elevation (ft)		WSEL (ft)		
(mi)	South	North	Existing	900-Ft	1,800-Ft
3.25	82.91	45.83	40.18	39.54	39.21
3.33333	77.77	46.02	40.31	39.69	39.36
3.41666	72.64	46.21	40.44	39.83	39.52
3.5	67.5	46.35	40.58	39.98	39.67
3.5715	55.94	46.34	40.71	40.12	39.82
3.643	48.55	47.28	40.83	40.26	39.97
3.644	48.9	47.28	40.84	40.27	39.97
3.65	48.54	47.1	40.87	40.3	40
3.651	48.54	47.1	40.87	40.3	40
3.7005	48.61	46.68	40.77	40.19	39.9
3.75	48.78	46.7	40.95	40.39	40.09
3.8215	48.69	49.88	41.04	40.5	40.21
3.893	50.3	50.9	41.17	40.62	40.34
3.894	49.3	56.97	41.17	40.62	40.34
3.913	53.99	50.39	41.28	40.74	40.46
3.914	55.96	56.96	41.29	40.75	40.47
4	48.96	47.8	41.35	40.81	40.54
4.08333	49.01	47.89	41.48	40.95	40.67
4.16666	49.27	48.44	41.61	41.09	40.82
4.25	49.13	53.05	41.75	41.24	40.98
4.33333	48.98	49.79	41.85	41.35	41.09
4.41666	48.84	50.43	41.97	41.48	41.23
4.5	48.69	52.09	42.09	41.61	41.37
4.58333	48.55	51.78	42.18	41.71	41.47
4.66666	48.66	51.54	42.27	41.8	41.57
4.75	49.09	50.13	42.36	41.9	41.67
4.83333	49.53	48.72	42.45	41.99	41.77
4.91666	49.88	47.66	42.55	42.11	41.88
5	50.14	48.15	42.66	42.23	42.01
5.08333	50.4	46.47	42.78	42.35	42.14
5.16666	50.64	45.91	42.93	42.52	42.31
5.25	50.69	46.02	43.16	42.76	42.56
5.33333	50.74	46.13	43.36	42.98	42.79

Rvr Sta	Levee Elevation (ft)		WSEL (ft)		
(mi)	South	North	Existing	900-Ft	1,800-Ft
5.41666	50.79	46.24	43.57	43.21	43.02
5.5	51.25	46.62	43.81	43.45	43.27
5.58333	50.99	46.61	43.9	43.55	43.38
5.66666	51.1	46.79	44	43.65	43.48
5.75	51.47	46.59	44.12	43.77	43.6
5.83333	51.85	46.36	44.05	43.71	43.54
5.91666	51.88	46.44	43.9	43.55	43.38
6	51.64	46.66	43.64	43.28	43.11
6.08333	51.45	46.88	43.94	43.61	43.44
6.16666	51.33	47.1	44.42	44.11	43.96
6.25	51.51	47.5	45.05	44.77	44.64
6.32	51.67	47.46	45.55	45.3	45.18
6.39	51.84	47.65	45.97	45.74	45.63
6.45	52.4	47.81	46.34	46.12	46.01
6.461	53.87	47.82	46.36	46.14	46.04
6.471	53.48	47.84	46.44	46.23	46.12
6.472	52	47.99	46.45	46.23	46.13
6.5	52.06	47.9	46.5	46.29	46.19
6.58333	52.24	48.09	46.76	46.55	46.45
6.66666	52.54	48.39	46.99	46.79	46.69
6.75	52.99	48.83	47.25	47.06	46.97
6.81566	53.35	49.17	47.47	47.28	47.19
6.88133	53.6	50.74	47.72	47.54	47.45
6.947	53.38	53.16	47.98	47.81	47.72
6.948	53.38	53.21	48.03	47.85	47.77
6.951	53.4	53.4	48.15	47.97	47.88
6.952	53.37	52.32	48.1	47.92	47.83
7	53.22	50.11	48.37	48.2	48.11
7.08333	52.93	50.42	48.91	48.75	48.67
7.16666	52.88	50.75	49.34	49.18	49.11
7.25	52.88	51.05	49.66	49.51	49.43
7.33333	53.21	51.37	49.72	49.57	49.5
7.41666	53.8	51.68	49.85	49.7	49.63
7.5	54.4	52	50	49.86	49.79
7.5555	55.48	52.21	50.2	50.06	49.99
7.611	58.62	54.5	50.38	50.24	50.17
7.613	59.33	56.58	50.46	50.32	50.25
7.62	59.33	56.58	50.57	50.4	50.32
7.621	59.5	56.69	50.5	50.33	50.24
7.623	59.5	58.7	50.49	50.32	50.24
7.624	54.63	52.51	50.53	50.36	50.27
7.631	54.6	52.54	50.72	50.52	50.42
7.632	58.35	55.9	50.66	50.46	50.36
7.691	54.3	54.22	50.81	50.61	50.51
7.75	54	53.06	50.94	50.74	50.65
7.83333	53.58	53.44	51.2	51	50.91
7.91666	53.65	53.72	51.37	51.18	51.09
8	54.22	54.16	51.52	51.33	51.24
8.08333	54.02	54.26	51.56	51.37	51.28
8.16666	54.21	54.53	51.58	51.39	51.3
8.25	54.7	54.83	51.56	51.37	51.28
8.33333	54.62	55.07	51.63	51.45	51.36
8.41666	54.9	55.34	51.72	51.54	51.45
8.5	55.2	55.61	51.85	51.67	51.59
8.58333	55.49	55.89	52.02	51.84	51.76
8.66666	55.76	55.94	52.2	52.03	51.95
8.75	56.03	56	52.43	52.26	52.18
8.83333	58.09	56.06	52.71	52.55	52.47
8.91666	56.07	56.15	53.01	52.85	52.77
9	56.05	56.25	53.33	53.07	53
9.0655	57.97	57.48	53.14	52.98	52.91
9.131	60.39	60.25	52.98	52.82	52.75
9.132	59.19	60.08	52.94	52.78	52.71
9.146	55.97	56.62	52.95	52.78	52.71
9.147	58.97	60.04	52.97	52.82	52.74
9.1985	55.95	58.72	53.06	52.9	52.83
9.25	57.23	57.63	53.14	52.99	52.91
9.33333	56.31	56.71	53.35	53.2	53.12
9.41666	56.79	56.97	53.54	53.39	53.32
9.5	57.81	57.25	53.8	53.66	53.59
9.58333	58.32	57.36	54.02	53.88	53.81
9.66666	58.83	57.47	54.26	54.13	54.06
9.75	59.38	57.53	54.42	54.28	54.22
9.83333	58.22	57.64	54.6	54.47	54.41
9.91666	58.47	57.75	54.7	54.57	54.51
10	60.4	58.26	54.84	54.71	54.65
10.0833	60.54	58.18	55.56	55.45	55.39
10.1666	60.69	58.39	55.98	55.87	55.81
10.25	60.88	58.61	56.29	56.18	56.13
10.3333	60.47	58.76	56.51	56.4	56.35
10.4166	60.88	58.93	56.68	56.57	56.52
10.5	61.3	59.39	56.81	56.71	56.66
10.555	61.36	59.26	56.88	56.77	56.72
10.61	61.42	59.43	56.94	56.84	56.79
10.68	61.5	59.63	57.03	56.93	56.88
10.75	61.58	59.84	57.11	57.01	56.96
10.8333	61.81	60.18	57.16	57.06	57.01
10.9166	62.08	60.55	57.19	57.09	57.04
11	57.34	60.83	57.19	57.09	57.04
11.0833	57.45	60.98	57.22	57.12	57.07
11.1666	57.51	61.08	57.19	57.09	57.04
11.25	57.57	61.2	57.12	57.02	56.98

Table A-15: Sacramento Bypass - North Levee Freeboard: Existing Conditions

River Station (mi)	Freeboard (ft)		1957 Design	
	100-Year Event	200-Year Event	River Station (mi)	Freeboard (ft)
0	4.27	2.97	0	6.73
0.000183	4.27	2.97	0.5	6.32
0.18	2.48	1.17	1	6.85
0.37	3.16	1.8	1.49	7.00
0.56	4.07	2.65	1.68	6.40
0.75	4.26	2.78		
0.93	4.55	3.01		
1.12	3.53	1.93		
1.3	5.03	3.39		
1.49	5.12	3.42		
1.67981	7.25	5.48		
1.68	6.62	4.65		

Table A-16: Sacramento Bypass - South Levee Freeboard: Existing Conditions

River Station (mi)	Freeboard (ft)		1957 Design	
	100-Year Event	200-Year Event	River Station (mi)	Freeboard (ft)
0			0	
0.000183			0.5	5.61
0.18	3.26	1.95	1	6.54
0.37	2.81	1.45	1.49	5.42
0.56	3.19	1.77	1.68	5.16
0.75	5.54	4.06		
0.93	3.47	1.93		
1.12	4.54	2.94		
1.3	3.86	2.22		
1.49	3.54	1.84		
1.67981	6.01	4.24		
1.68	5.38	3.41		

Table A-17: Sacramento Bypass - North Levee Freeboard: 900 Foot Expansion

River Station (mi)	Freeboard (ft)		1957 Design	
	100-Year Event	200-Year Event	River Station (mi)	Freeboard (ft)
0	4.17	2.84	0	6.73
0.000183	4.17	2.84	0.5	6.32
0.18	2.39	1.05	1	6.85
0.37	3.19	1.82	1.49	7.00
0.56	4.19	2.78	1.68	6.40
0.75	4.46	3.03		
0.93	4.88	3.41		
1.12	3.97	2.47		
1.3	5.55	4.02		
1.49	5.74	4.17		
1.67981	8.03	6.42		
1.68	7.63	5.88		

Table A-18: Sacramento Bypass - South Levee Freeboard: 900 Foot Expansion

River Station (mi)	Freeboard (ft)		1957 Design	
	100-Year Event	200-Year Event	River Station (mi)	Freeboard (ft)
0			0	
0.000183			0.5	5.61
0.18	3.17	1.83	1	6.54
0.37	2.84	1.47	1.49	5.42
0.56	3.31	1.9	1.68	5.16
0.75	5.74	4.31		
0.93	3.8	2.33		
1.12	4.98	3.48		
1.3	4.38	2.85		
1.49	4.16	2.59		
1.67981	6.79	5.18		
1.68	6.39	4.64		

Table A-19: Sacramento Bypass - North Levee Freeboard: 1,800 Foot Expansion

River Station (mi)	Freeboard (ft)		1957 Design	
	100-Year Event	200-Year Event	River Station (mi)	Freeboard (ft)
0	4.11	2.76	0	6.73
0.000183	4.11	2.76	0.5	6.32
0.18	2.34	0.98	1	6.85
0.37	3.18	1.79	1.49	7.00
0.56	4.22	2.82	1.68	6.40
0.75	4.53	3.11		
0.93	4.98	3.55		
1.12	4.13	2.67		
1.3	5.74	4.27		
1.49	6	4.5		
1.67981	8.36	6.84		
1.68	8.09	6.47		

Table A-20: Sacramento Bypass - South Levee Freeboard: 1,800 Foot Expansion

River Station (mi)	Freeboard (ft)		1957 Design	
	100-Year Event	200-Year Event	River Station (mi)	Freeboard (ft)
0			0	
0.000183			0.5	5.61
0.18	3.12	1.76	1	6.54
0.37	2.83	1.44	1.49	5.42
0.56	3.34	1.94	1.68	5.16
0.75	5.81	4.39		
0.93	3.9	2.47		
1.12	5.14	3.68		
1.3	4.57	3.1		
1.49	4.42	2.92		
1.67981	7.12	5.6		
1.68	6.85	5.23		

APPENDIX B. SAMPLE CALCULATIONS

Weir capacity calcs

The project design capacity of the weir is 112,000 cfs

The existing weir has 48 gates

Capacity per gate = $\frac{112,000 \text{ cfs}}{48 \text{ gates}} = 2333.3 \frac{\text{cfs}}{\text{gate}}$

Capacity of the new weir (900 ft Bypass extension)

$= 2,333.3 \frac{\text{cfs}}{\text{gate}} \times 72 \text{ gates} = 168,000 \text{ cfs}$

Capacity of the new weir (1800 ft Bypass extension)

$= 2,333.33 \frac{\text{cfs}}{\text{gate}} \times 94 \text{ full size gates} + 2,333.33 \left(\frac{21.6 \text{ ft}}{38.25 \text{ ft}}\right) =$

$= 220,551 \text{ cfs}$

proportion of a 95th gate to the typical gate

APPENDIX C. ADDITIONAL MAPS AND FIGURES

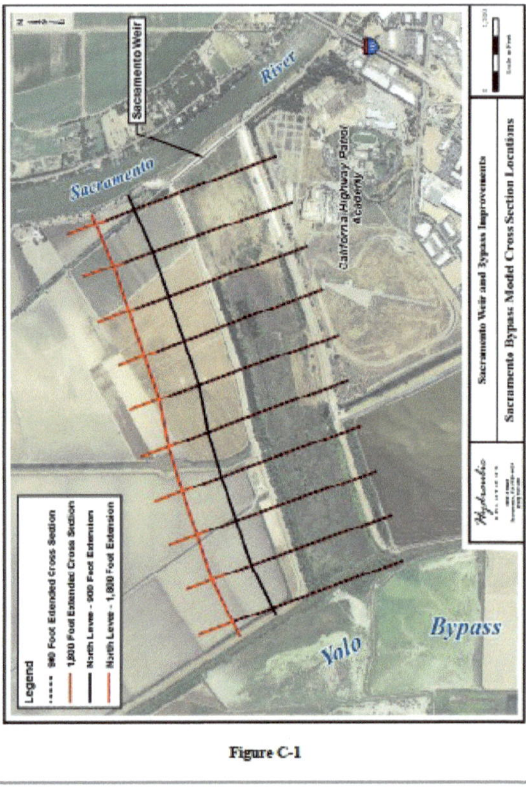

Figure C-1

Type of Channel and Description	Minimum	Normal	Maximum
A. *Natural Streams*			
1. **Main Channels**			
a. Clean, straight, full, no rifts or deep pools	0.025	0.030	0.033
b. Same as above, but more stones and weeds	0.030	0.035	0.040
c. Clean, winding, some pools and shoals	0.033	0.040	0.045
d. Same as above, but some weeds and stones	0.035	0.045	0.050
e. Same as above, lower stages, more ineffective slopes and sections	0.040	0.048	0.055
f. Same as "d" but more stones	0.045	0.050	0.060
g. Sluggish reaches, weedy, deep pools	0.050	0.070	0.080
h. Very weedy reaches, deep pools, or floodways with heavy stands of timber and brush	0.070	0.100	0.150
2. **Flood Plains**			
a. Pasture no brush			
1. Short grass	0.025	0.030	0.035
2. High grass	0.030	0.035	0.050
b. Cultivated areas			
1. No crop	0.020	0.030	0.040
2. Mature row crops	0.025	0.035	0.045
3. Mature field crops	0.030	0.040	0.050
c. Brush			
1. Scattered brush, heavy weeds	0.035	0.050	0.070
2. Light brush and trees, in winter	0.035	0.050	0.060
3. Light brush and trees, in summer	0.040	0.060	0.080
4. Medium to dense brush, in winter	0.045	0.070	0.110
5. Medium to dense brush, in summer	0.070	0.100	0.160
d. Trees			
1. Cleared land with tree stumps, no sprouts	0.030	0.040	0.050
2. Same as above, but heavy sprouts	0.050	0.060	0.080
3. Heavy stand of timber, few down trees, little undergrowth, flow below branches	0.080	0.100	0.120
4. Same as above, but with flow into branches	0.100	0.120	0.160
5. Dense willows, summer, straight	0.110	0.150	0.200

Figure C-2: HEC-RAS Reference Manual "n" values

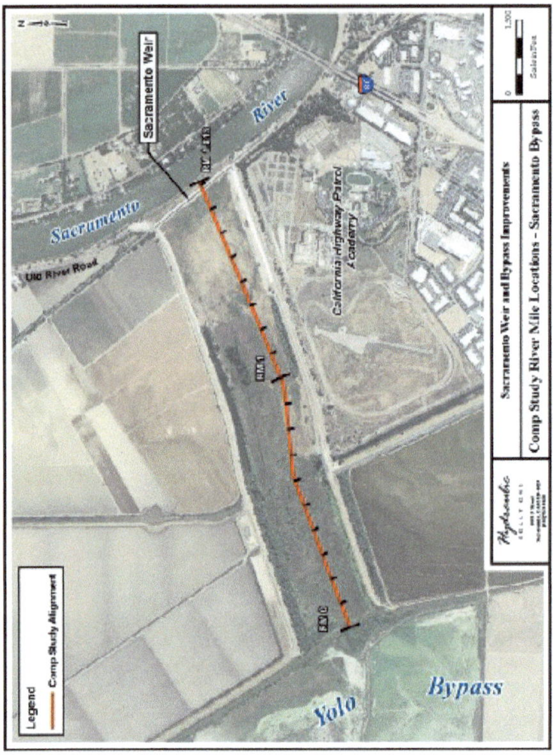
Sacramento Weir
Sacramento
River
California Highway Patrol Academy
Yolo
Bypass
Legend
Comp Study Alignment
Sacramento Weir and Bypass Improvements
Comp Study River Mile Locations – Sacramento Bypass

Figure C-3

Knights Landing
Woodland
Yolo
Bypass
West
Sacramento
Davis
Legend

Figure C-4

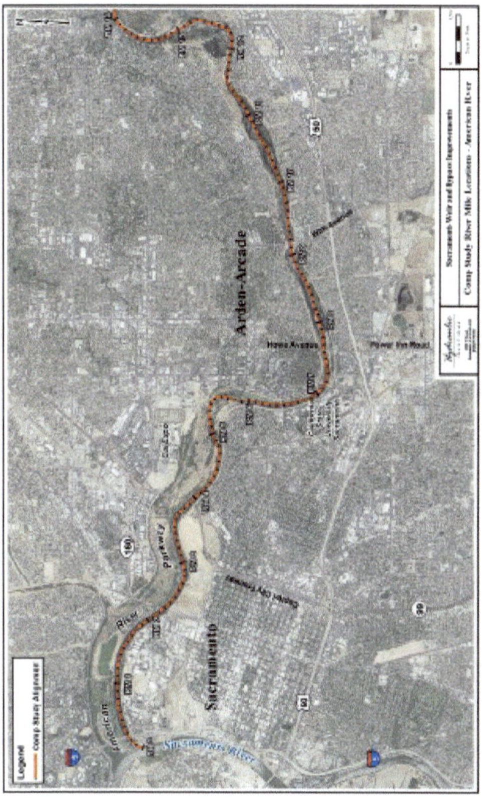
Legend
Comp Study Alignment
Arden-Arcade
Sacramento
American
River
Parkway
Howe Avenue
Power Inn Road
Sacramento River
Sacramento Weir and Bypass Improvements
Comp Study River Mile Locations - American River

Figure C-5

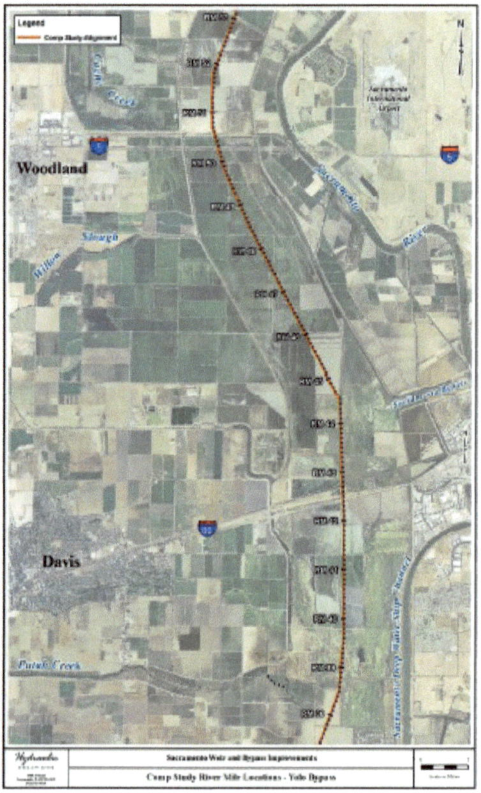
Legend
Woodland
Davis
Willow
Slough
Putah Creek
Sacramento Weir and Bypass Improvements
Comp Study River Mile Locations - Yolo Bypass

Figure C-6

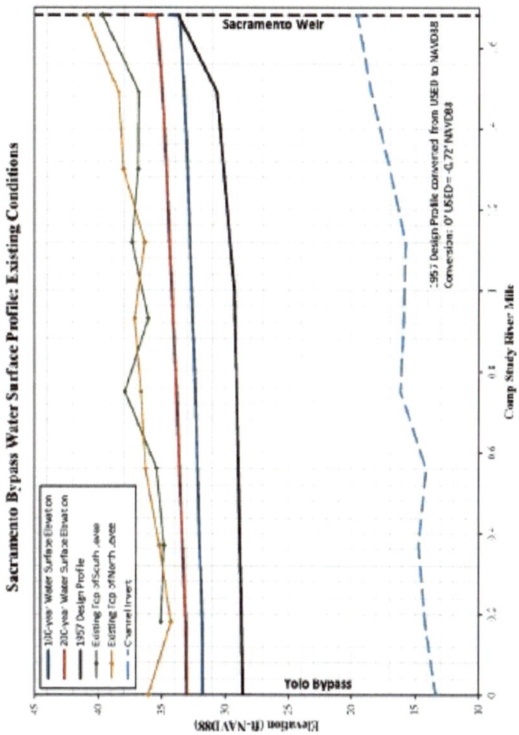

Figure C-7: Comparison of 100-year and 200-year WSEL: Existing Conditions

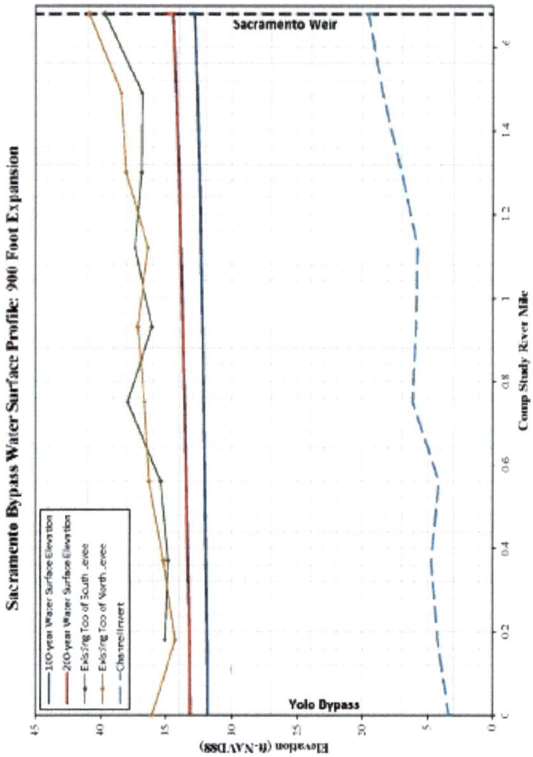

Figure C-8: Comparison of 100-year and 200-year WSEL: 900 Foot Expansion

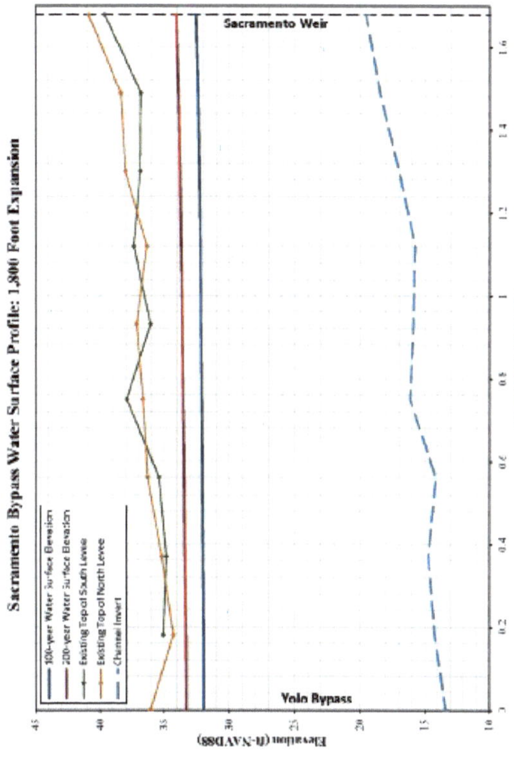

Figure C-9: Comparison of 100-year and 200-year WSEL: 1,800 Foot Expansion

ABOUT THE AUTHOR

Grace Lau graduated civil engineering school in 2013. She is currently an engineering intern in San Francisco. She likes to volunteer, hang out with friends, and read.

www.ingramcontent.com/pod-product-compliance
Lightning Source LLC
Chambersburg PA
CBHW040807200526
45159CB00022B/42
9781497540453